Evolution:

A Theory in Crisis

Evolution:
A Theory in Crisis

Michael Denton

ADLER&ADLER

Published in the United States in 1986 by
Adler & Adler, Publishers, Inc.
4550 Montgomery Avenue
Bethesda, Maryland 20814

Originally published in Great Britain by
Burnett Books Limited
Produced and Distributed by
The Hutchinson Publishing Group

Library of Congress Cataloging in Publication Data

Denton, Michael.
Evolution: a theory in crisis.
Reprint. Originally published: Great Britain:
Burnett Books, 1985.
1. Evolution.   I.   Title.
QH371.D46   1986   575   85–13556
ISBN 0–917561–05–8

Printed in the United States of America

10   9   8   7   6   5

Contents

	List of Illustrations	7
	Acknowledgements	9
	Preface	15
1	Genesis Rejected	17
2	The Theory of Evolution	37
3	From Darwin to Dogma	69
4	A Partial Truth	79
5	The Typological Perception of Nature	93
6	The Systema Naturae from Aristotle to Cladistics	119
7	The Failure of Homology	142
8	The Fossil Record	157
9	Bridging the Gaps	199
10	The Molecular Biological Revolution	233
11	The Enigma of Life's Origin	249
12	A Biochemical Echo of Typology	274
13	Beyond the Reach of Chance	308
14	The Puzzle of Perfection	326
15	The Priority of the Paradigm	344
	Index	360

List of Illustrations

1.1	The Galapagos Finches	31
2.1	The First Evolutionary Tree	38
2.2	The Phenomenon of Homology	49
5.1	Variations on an Invariant Theme	95
5.2	The Heart and Aortic Arches in the Major Vertebrate Types	112
5.3	The Arrangement of the Heart and Aortic Arches in the Amphibia, Reptiles and Mammals	114
5.4	Early Embryology in Amphibians, Reptiles and Mammals	115
6.1	The Classification of the Vertebrates According to M. Milne Edwards (1844)	120
6.2	Four Evolutionary Trees	126–7
6.3	The Typological Perception of Nature	133
7.1	The Basic Pentadactyl Design	142
7.2	A Pleiotropic Gene in the Domestic Fowl	150
8.1	The Coelacanth	157
8.2	Discontinuities in the Fossil Record	167–71

     First Amphibian – *Ichthyostega*
     Rhipidistian fish – *Eusthenopteron*
     Pterosaur – *Dimorphodon*
     Archaeopteryx
     Euparkeria
     First bat – *Icaronycteris*
     Modern Shrew
     Mesosaurus
     Ichthyosaur
     Hylonomus
     Plesiosaur – *Crytocleidus*
     Araeocelis
     Early whale – *Zygorhiza kochi*
     Sinopia
     Seal

Cynodictis
Sirenian – *Halitherium*
Hyrax – *Procavia*

8.3	Adaptive Radiation of Vertebrates	173
8.4	The Evolution of the Horse	183
8.5	The Adequacy of the Fossil Record	190
9.1	Pro-avis	200
9.2	The Parabronchi of the Avian Lung	211
9.3	The Rotary Motor of the Bacterial Flagellum	224
10.1	The Chemical Structure of Three Amino Acids	235
10.2	The Chemical Structure of a Short Section of the Amino Acid Chain of a Protein	236
10.3	The Structural Organization of a Protein	237
10.4	The Structure of DNA	241
10.5	The Elongation of the Amino Acid Chain	246
10.6	DNA Replication	247
12.1	The Cytochromes Percent Sequence Difference Matrix	279
12.2	The Molecular Equidistance of all Eucaryotic Organisms from Bacteria	280
12.3	Mutation Rates per Unit Time	299

Acknowledgements

I am grateful to the following for permission to reproduce copyright material:

To Academic Press and Professor J. Beuttner-Janusch for material from *Evolutionary and Genetic Biology of Primates* ed by J. Beuttner Janusch (1963). To Academic Press for permission to reproduce the figure of *Icaronycteris index* from *Biology of Bats*, ed W. A. Wimsatt (1970). To Academic Press and Professor Harry J. Jerison for material from *Evolution of the Brain and Intelligence* (1973). To the American Association for the Advancement of Science for material from the article "Biological Classification: Toward a Synthesis of Opposing Methodologies" by E. Mayr in *Science* 214, pp 510–516, copyright 1981 by the American Association for the Advancement of Science; for material from the article "The Nature of the Darwinian Revolution" by E. Mayr in *Science* 176, pp 981–989, copyright 1972 by the American Association for the Advancement of Science; and the material from the article "The Problem of Plan and Purpose in Nature" by G. G. Simpson in *Scientific Monthly*, 64, pp 481–495 (1947). To the American Museum of Natural History for permission to reproduce the figures of *Coelacanth latimeria, Crytocleidus oxoniensis, Crocidura, Sinopa rapax, Cynodictis gregarius, Zygorhiza kochii, Procavia*, and the common seal from *Evolution Emerging*, Vol II, by W. K. Gregory (1951). To the *American Scientist* for material from the article "Bird Flight: How did it Begin?" by John Ostrom in *American Scientist*, 67, pp 46–56 (1979) and to John Ostrom for kindly providing me with a glossy print of Pro-avis. To the American Society of Biological Chemists for permission to reproduce the "Illustration of the Amino Acid Sequence of Bovine Ribonuclease" from the article "The Sequence of Amino Acid Residues in Bovine Ribonuclease" by D. G. Smyth, W. H. Stein, and S. Moore, *J. of Biol. Chem.*, 238, pp 227–234, (1963). To Lady Nora Barlow for material from *Autobiography of Charles Darwin* published by Collins, London, 1958. To Cambridge University Press for permission to reproduce some of the illustrations of Darwin's Finches from *Darwin's Finches* by David Lack (1947); for material from *On Growth and Form* by D'Arcy Thompson (1942); and for material from *The Biology of the Dragonfly* by R. J. Tillyard (1917). To the University of Chicago Press for permission to quote and to reproduce the following illustrations: *Mesosaurus*, Ichthyosaur, *Halitherium*, from *Vertebrate Paleontology* by A. S. Romer, third edition (1966); and for material from the article "The History of Life" by G. G. Simpson in *Evolution of Life*, ed Sol Tax (1960). To William Collins and Sons and Sir Alister Hardy for permission to reproduce the illustration of homology in the design of vertebrate forelimbs from *The Living Stream* by A. Hardy (1965). To Robert G. Colodny for material from the article "Problems of Empiricism" by P. Feyerabend in *Beyond the Edge of Certainty*, ed R. G. Colodny (1965). To Columbia University Press for material from *Evolution Above the Species Level* by Bernhard Rensch (1959). To Dr H. R. Duncker of the Zentrüm für Anatomie und Cytobiologie der Justus Liebig Universität and to Springer-Verlag for permission to reproduce the illustration of the parabronchi of the avian lung. To Editions du Seuil for permission to reproduce the illustrations of the structure and replication of DNA from *Le Hasard et La Nécessité* by Jacques Monod (1970). To Elsevier Scientific Publishing Company and J. W. Valentine for permitting me to reproduce the illustration of "4 Phylogenies of Metazoan Phyla" from the article "General Patterns of Metazoan Evolution" by J. W. Valentine in *Patterns of Evolution as Illustrated by the Fossil Record*, ed A. Hallam (1977). To the Encyclopaedia Britannica for material from "Evolution" in Encyclopaedia Britannica, 15th edition (1974) 7: 8–9. To Peter Forey and the Royal Society of London for permission to quote material from the Proceedings of the Royal Society. To W. H. Freeman and Company for the use of material from the article "How Bacteria Swim" by H. C. Berg, *Scientific American*, August 1975; for material from the article "Artificial Intelligence" by M. L. Minsky, *Scientific American*, September 1966; and from "Darwin's Missing Evidence" by H. B. D. Kettlewell, *Scientific American*, March 1959. To Harvard University Press for the use of material from *The Great Chain of Being* by A. O. Lovejoy (1951) and from *Georges Cuvier, Zoologist* by W. Coleman (1964). To Professor David L. Hull and Harvard University Press for material from *Darwin and his Critics* (1973). To Holden Day Inc. for permission to quote from *Intelligent Life in the Universe* by C. Sagan (1978). To the Hutchinson Publishing Group for the use of material from *Beyond Reductionism* ed A. Koestler (1969). To the University of Illinois Press for material from *Phylogenetic Systematics* by W. Hennig (1966). To Professor Erik Jarvik for permission to copy his reconstruction of the skeleton of *Ichthyostega*. To the Linnean Society of London for permission to copy the illustration of *Hylonomus*. To the Longman Group Ltd for permission to use material from *Organisation and Evolution in Plants* by C. W. Wardlaw (1965); and from *Social History of England* by G. M. Trevelyan (1944). To Professor Ernst Mayr for the use of material from his book *Population, Species and Evolution* published by Harvard University Press (1970). To the McGraw-Hill Book Company for material from *Methods and Principles of Systematic*

Zoology, by E. Mayr et al (1953); and permission to reproduce the illustration of horse evolution from *Elements of Zoology*, 4th edition, by T. I. Storer et al (1977). To C. V. Mosby Company and G. C. Kent for permission to reproduce the illustration of "The Vertebrate Heart and Aortic Arches" from *Comparative Anatomy of the Vertebrates*, 5th edition, by G. C. Kent (1983). To the Museum of Comparative Zoology, Harvard University, for permission to copy the figure of *Araeoscelis*. To Neale Watson Academic Publications Inc and Professor Martin Rudwick for material from *The Meaning of the Fossils* (1972). To the New York Academy of Science for permission to copy the illustration of *Eusthenopteron*. To Oxford University Press for the use of material from *Homology: An Unsolved Problem*, by G. R. De Beer (1971). To W. B. Saunders Company Ltd for permission to reproduce two illustrations, "The Adaptive Radiation of the Mammals" and "The Vertebrate Heart and Aortic Arches" from *The Vertebrate Body* by A. S. Romer, 5th edition (1977). To the Royal Society of London for permission to copy the illustration of *Euparkeria* and to Professor Barbara Stahl for providing the glossy print. To Hobart M. Smith for permission to reproduce the illustration of the pentadactyl design from his book *Evolution of Chordate Structure*, published by Holt, Rinehart and Winston Inc (1960). To Temple University Press for use of material from *Charles Darwin: The Years of Controversy* by P. J. Vorzimmer (1970). To Georg Thieme Verlag for permission to reproduce the illustration of "A Pleiotropic Gene in the Domestic Fowl" from *Developmental Genetics and Lethal Factors*, by E. Hadorn (1961). To Springer-Verlag for use of material from the article "The Appearance of New Structures and Functions in Proteins During Evolution", by E. Zuckerkandl, *J. Mol. Evol.*, 7, 1–57 (1975). To John Wiley and Sons, Inc Publishers, for permission to reproduce the illustration of "Principal Types of Chordate Cleavage" from *Analysis of Vertebrate Structure*, by M. Hildebrand (1974); for the use of material from the article "The Search for Extraterrestrial Technology" by J. Freeman Dyson in *Perspectives in Modern Physics*, ed R. E. Marshak (1966); and for permission to reproduce the illustration of feather structure from *Fundamentals of Ornithology*, by J. Van Tyne and A. J. Berger (1959). To Yale University Press for material from the article "Pre-Metazoan Evolution and the Origins of the Metazoa" by Preston E. Cloud, in *Evolution and Environment* ed Ellen T. Drake (1968).

The quotes at the beginning of each chapter are from Charles Darwin.

Nature, inexhaustible in fecundity and omnipotent in its works ... has been settled in the innumerable combinations of organic forms and functions which compose the animal kingdom by physiological incompatibilities alone. It has realized all those combinations which are not incoherent and it is these incompatibilities, this impossibility of the coexistence of one modification with another which establish between the diverse groups of organisms those separations, those gaps, which mark their necessary limits and which create the natural embranchments, classes, orders, and families.

Georges Cuvier, 1835

I can see no limit to the amount of change to organic beings which may have been affected in the long course of time through nature's power of selection.

Charles Darwin, 1859

Preface

The question of evolution is generating more controversy and argument today than at any other time since the "Great Debate" in the nineteenth century. At prestigious international symposia, in the pages of leading scientific journals and even in the sober galleries of the British Natural History Museum, every aspect of evolution theory is being debated with an intensity which has rarely been seen recently in any other branch of science.

It is not hard to understand why the question of evolution should attract such attention. The idea has come to touch every aspect of modern thought; and no other theory in recent times has done more to mould the way we view ourselves and our relationship to the world around us. The acceptance of the idea one hundred years ago initiated an intellectual revolution more significant and far reaching than even the Copernican and Newtonian revolutions in the sixteenth and seventeenth centuries.

The triumph of evolution meant the end of the traditional belief in the world as a purposeful created order – the so-called teleological outlook which had been predominant in the western world for two millennia. According to Darwin, all the design, order and complexity of life and the eerie purposefulness of living systems were the result of a simple blind random process – natural selection. Before Darwin, men had believed a providential intelligence had imposed its mysterious design upon nature, but now chance ruled supreme. God's will was replaced by the capriciousness of a roulette wheel. The break with the past was complete.

Because of its influence on fields far removed from biology, the current problems in evolution theory have been widely publicized and have captivated the public imagination to the extent that topics such as the gaps in the fossil record or competing methodologies in taxonomy – subjects which would normally be considered obscure and esoteric – are discussed in detail in popular magazines and even the daily press. Any suggestion that there might be something seriously wrong with the Darwinian view of nature is bound to excite public attention, for if biologists cannot substantiate the fundamental claims of Darwinism, upon which rests so much of the fabric of twentieth-century thought, then clearly the intellectual and philosophical implications are immense. Small wonder, then, that the current tumult in biology is arousing such widespread interest.

Basically there are two different philosophical approaches to the debate. On the one hand, one can adopt the conservative position and view the difficulties as essentially trivial, merely puzzling anomalies, that will all be eventually reconciled somehow to the traditional framework. Alternatively, one can adopt a radical position and view the problems not as puzzles, but as counterinstances or paradoxes which will never be adequately explained within the orthodox framework, and indicative therefore of something fundamentally wrong with the currently accepted view of evolution.

While most evolutionary biologists who have written recently about evolution concede that the problems are serious, nearly all take an ultimately conservative stand, believing that they can be explained away by making only minor adjustments to the Darwinian framework.

In this book I have adopted the radical approach. By presenting a systematic critique of the current Darwinian model, ranging from paleontology to molecular biology, I have tried to show why I believe that the problems are too severe and too intractable to offer any hope of resolution in terms of the orthodox Darwinian framework, and that consequently the conservative view is no longer tenable.

CHAPTER 1
Genesis Rejected

After having been twice driven back by heavy south-western gales,
Her Majesty's ship *Beagle*, a ten-gun brig, under the command of
Captain FitzRoy, RN, sailed from Devonport on 27 December 1831,
The object of the expedition was to complete the survey of Patagonia
and Tierra del Fuego, commenced under Captain King in 1826 to
1830; to survey the shores of Chile, Peru, and of some islands in the
Pacific; and to carry a chain of chronometrical measurements round
the world.

As the *Beagle* sailed out of Devonport in December 1831, it could
hardly have seemed to those on board that there was anything out of
the ordinary or fateful about the voyage ahead. Yet the observations
that Darwin was to make during his five years aboard "that good
little vessel", as he affectionately referred to her, were to sow in his
mind the seed of the idea of organic evolution. This was a seed which
was ultimately to flower in *The Origin of Species* into a new and
revolutionary view of the living world which implied that all the
diversity of life on Earth had resulted from natural and random
processes and not, as was previously believed, from the creative
activity of God. The acceptance of this great claim and the consequent elimination of God from nature was to play a decisive role
in the secularization of western society. The voyage on the *Beagle*
was therefore a journey of awesome significance. Its object was to
survey Patagonia; its result was to shake the foundations of western
thought.

The philosophy of nature held by Darwin, as he set sail on the
Beagle, and by most of his contemporaries, was completely antithetical to the idea of organic evolution. Biology in the early decades
of the nineteenth century was dominated by the idea that the organic
world was a fundamentally discontinuous system in which all the

major groups of organisms were unique and isolated and unlinked by transitional forms. Species were held to breed true to type, generation after generation, without ever undergoing any significant sort of change. Where there was variation, it was only trivial variation within the clearly defined limits of the species or type. Thus to the naturalists of the nineteenth century the basic order of nature was static and discontinuous, very different from the dynamic continuous model which was later to become axiomatic for most biologists after 1859.

The so-called typological model of nature adhered to by biologists early in the century was not without a considerable degree of empirical support. To anyone observing nature over a short span of time it must have seemed self evident that species bred to type generation after generation, and that the living world flowed according to a fixed and preordained plan. Moreover, the work of the great nineteenth-century comparative anatomists such as Cuvier and, later, Owen had shown that the living world could be considered divided into distinct types or phyla and that organisms clearly intermediate between different classes were virtually unknown.

Comparative anatomy had also revealed that organisms were integrated wholes in which all the components were coadapted to function together; and this seemed to many to preclude any sort of major evolutionary transformation. As William Coleman, an authority on Georges Cuvier, points out:[1]

> The organism, being a functionally integrated whole each part of which stood in close relation to every other part, could not, under pain of almost immediate extinction, depart significantly from the norms established for the species by the first anatomical rule.
>
> A major change, for example, a sharp increase in the heart beat or the diminution by half of the kidney and thus a reduction in renal secretion, would by itself have wrought havoc with the general constitution of the animal. In order that an animal might persist after a change of this magnitude it would be necessary that the other organs of the body be also proportionally modified. In other words, an organism much change en bloc or not at all. Only saltatory modification could occur, and this idea was to Cuvier, as it is to most modern zoologists, but for very different reasons, unverified and basically absurd. Transmutation by the accumulation of alterations, great or small, would thus be impossible.

The doctrine of the fixity of species was also derived from a great metaphysical system of thought known as essentialism which presumed according to Mayr:[2]

> . . . that the changeable world of appearances is based on underlying immutable essences, and that all members of a class represent the same essence. This idea was first clearly enunciated in Plato's concept of the eidos.

hence

> The observed vast variability of the world has no more reality, according to this philosophy, than the shadows of an object on a cave wall, as Plato expressed it in his allegory. The only things that are permanent, real, and sharply discontinuous from each other are the fixed, unchangeable "ideas" underlying the observed variability. Discontinuity and fixity are, according to the essentialist, as much the properties of the living as of the inanimate world.

In other words, all individual entities were physical expressions of a finite number of ideal unchanging forms. Applied to the biological sphere, it followed that there were fixed bounds determined by the form of the underlying type beyond which biological variation could not go: nature was, therefore, fundamentally discontinuous.

Typological thinking and the idea of the fixity of species can be traced back in biological thought to Aristotle, who in turn had derived it from the Platonic doctrine of the *eidos*. For centuries during the middle ages Aristotle's philosophy of nature had been the official doctrine of the Church and his views still carried great weight among biologists of the early nineteenth century.

The discontinuous typological view of nature was also sanctioned to some extent by religious belief. The religious climate in England when the *Beagle* sailed in 1831 was very different to that of today. Scientific knowledge was not looked upon as a challenge to religious belief as it was to become after the acceptance of evolution. The discoveries of science, and particularly the magnificently ordered vision of the physical world implicit in the Newtonian synthesis, were all taken as evidence for the existence of a creator and the grandeur of his design. The enormous appeal of natural theology, epitomised by the popularity of William Paley and his famous *Evi-*

dences, illustrates how widespread was the view that science supported theological claims.

An indication of how prevalent such thinking was, even among members of the scientific community, is seen in the opening issue of the prestigious *Zoological Journal of London*, founded in 1824. The editor's introduction in this first issue emphasizes the idea that the study of nature reveals the wisdom of God and indicates the special place occupied by man in the natural order of the world:[3]

> The naturalist . . . sees the beautiful connection that subsists through-
> out the whole scheme of animated nature. He traces . . . a mutual
> depending that convinces him nothing is made in vain. He feels, too,
> that at the head of all this system of order and beauty, pre-eminent in
> the domain of his reason, stands Man . . . the favoured creature of his
> Creator.

In the same spirit in 1857, only two years before Darwin's *Origin*, one of the leading biologists of North America, Louis Agassiz, at that time Professor of Zoology at Harvard, could write that the living world[4]

> . . . shows also premeditation, wisdom, greatness, prescience, omni-
> science, providence . . . all these facts . . . proclaim aloud the One God
> whom man may know, and natural history must, in good time become
> the analysis of the thoughts of the Creator of the Universe, as mani-
> fested in the animal and vegetable kingdoms, as well as in the inorganic
> world.

As far as Darwin's contemporaries were concerned, few felt anything of the conflict between science and religion which is so characteristic of twentieth century thought. The conflict between science and religion only erupted later in the nineteenth century when it became generally acknowledged that discoveries in geology and biology were incompatible with a literal Genesis. A man of science could still, in the first decades of the nineteenth century, accept the account of creation in the first book of the Bible as a literal historical description of the origin of the world. There was indeed an element of conflict between the miraculous suspension of natural law that Genesis implied and the increasing success of science in eliminating any need for supernatural explanations, but it was perceived to be minimal.

But although biblical literalism was not yet in open conflict with scientific knowledge it undoubtedly had a constraining effect on geological and biological thought throughout the eighteenth and the first part of the nineteenth centuries. Coleman has written of those times:[5]

> Many naturalists were strongly influenced by the seeming necessity of finding in nature the literal realization of the events catalogued in Divine Scripture. On many issues of natural history and particularly that of the nature of the biological species, it was commonly believed that the bible was to be either the final authority or at least a repository of general truths of which none could be safely or legitimately disregarded by a truly philosophical naturalist.

While it is difficult from this distance to judge just how much influence the belief in a literal Genesis had on the great naturalists of the first decades of the nineteenth century it was undoubtedly considerable.* For example, it was almost certainly the major factor responsible for the widespread belief in a six thousand-year-old earth; and the doctrine of the fixity of species seemed to be supported by statements in Genesis which appeared to imply that species had been created "after their kind" and bred "true to type".

But there were an increasing number of observations, particularly geological, that were difficult to reconcile with the Mosaic account. It was increasingly obvious to most geologists that none of the known natural processes, such as water or wind erosion, could have shaped the Earth's surface in a mere six thousand years. These processes cause virtually no perceptible change even over centuries, yet the book of Genesis implied that the Earth had been created in the relatively recent past, only six thousand years ago, according to some biblical chronologists.

Another challenge to the traditional account was the discovery that many species which had once lived on the Earth had become extinct. The biblical deluge could not have been responsible for the extinctions of all these ancient life forms because Genesis implied that every species was rescued from the flood. To complicate this problem there was some evidence that the Earth may have been populated by a whole succession of very different faunas.

*The degree to which religious belief influenced leading nineteenth-century biologists such as Cuvier, Owen, Lyell and others has probably been exaggerated; see discussion in Chapter Five.

In an attempt to reconcile this new knowledge with the biblical story, a compromise was proposed in the theory of catastrophes, which supposed that the history of the Earth had been periodically interrupted by great cataclysms of supernatural origin causing sudden massive changes to the surface of the Earth. Following each catastrophe it was believed that the Earth had been repopulated by newly created species.

As Cuvier expressed it:[6]

> Life in those times was often disturbed by these frightful events. Numberless living things were victims of such catastrophes: some, inhabitants of the dry land, were engulfed in deluges; others, living in the heart of the seas, were left stranded when the ocean floor was suddenly raised up again; and whole races were destroyed forever, leaving only a few relics which the naturalist can scarcely recognize.

Evidence for catastrophes was found in all kinds of geological phenomena: sea shells on mountain tops; vast alluvial beds of gravels and clays; the sudden extinction and subsequent freezing of such prehistoric species as the mammoth; massive heaps of bones of extinct species found in caves in various localities; large blocks of stone which clearly had been carried from distant sources and dropped in their present location. Even the carving out of river valleys was put down to the action of catastrophic deluges, and the thrusting up of mountain chains was similarly considered to have resulted from cataclysmic upheavals. Catastrophism was essential if the vast geological changes which must have occurred in the past were to be accounted for in the short time span implied by a literal interpretation of Genesis.

Catastrophism did not reconcile geology with Moses quite as satisfactorily as some might have wished. The Bible made no mention of catastrophes which could have eliminated prehistoric life preceding the creation of man, and it was difficult to see how such events could have occurred within the short space of six days. One way out of this dilemma was to propose, as Cuvier did, that each day was a period of indefinite length, thus allowing ample time for the occurrence of catastrophes. Another popular stratagem was to propose that the six days' creation was not an account of the original creation of the world but rather a description of God's activity in restoring and repopulating the Earth following the last great cataclysm which preceded the flood of Noah. According to Thomas Chalmers, a leading apologist of the period:[7]

> Moses may be supposed to give us not a history of the first formation of things, but of the formation of the present system; and as we have already proved the necessity of direct exercises of creative power to keep up the generations of living creatures; so Moses may, for anything we know, be giving us the full history of the last great interposition, and be describing the successive steps by which the mischiefs of the last catastrophe were repaired.

All in all the scientific community was still wedded to the cosmology of Genesis and to a belief in the literal historicity of Scripture. There were, of course, arguments over various details. Not everyone agreed as to how many catastrophes had occurred or which particular geological phenomena could be attributed to the flood of Noah and which to previous catastrophes. There were also disagreements as to whether new species had been created as an original pair or whether several pairs had been created simultaneously. Another source of discussion was the way in which different animal species had reached remote regions of the globe after the flood. The existence of these disputes again serves to illustrate the theological orientation of the natural sciences in this period and emphasizes the enormous intellectual gulf that separates their outlook from the secular ethos of today. It is extraordinary to think that a little over one hundred years ago the majority of biologists and geologists felt completely at ease intellectually with a view of the past which necessitated innumerable miraculous or supernatural events.

The consensus in favour of Genesis was not quite universal, however. There was a minority of dissenters who questioned the established position and whose views were, as it turned out, a foreshadowing of the intellectual upheaval that was shortly to come. The possibility of the transmutation of species had been considered by the French biologist Buffon in the mid-eighteenth century and later by another French biologist, Lamarck. Even Darwin's grandfather, Erasmus Darwin, had toyed with the idea of evolution in his major work *Zoonomia*.

Both catastrophism and the idea of a six-thousand-year-old Earth, were also being questioned by a minority of geologists. Already in the eighteenth century the Scottish geologist Hutton had presented an alternative uniformitarian interpretation of geological phenomena, arguing that the geological appearance of the Earth could be readily accounted for by small changes which were occurring continually, uninterrupted over vast periods of time. Shortly before Darwin set sail on the *Beagle* the pre-eminent exponent of uniformitarianism, Charles Lyell, published the first volume of his great work *Principles of Geology Being an Attempt to Explain the Former Changes of the Earth's Surface by Reference to Causes now in Operation*. Lyell's work ultimately proved a watershed in geological thought and convinced most scientists, including Darwin, of the validity of uniformitarianism; but at the time of publication it was still greatly considered heretical by the majority of geologists. Thus, despite the existence of such straws in the wind, the overwhelming consensus among biologists was in favour of catastrophism and a recent Earth, and the special creation of each and every species as a fundamentally immutable entity.

Darwin was born into a well-to-do middle class English family at the turn of the nineteenth century. His father was a Shropshire doctor who gave his son the typical education of his class. He was on his own admission 'rather below average' and more interested in shooting and sport than in drilling Greek and Latin verbs. From school Darwin went to study medicine at Edinburgh, at that time one of the leading medical schools in Europe, but the subject bored him, so his father, fearing that he might turn into an idle sporting man, sent him to Cambridge to study theology, intending him eventually to settle as a country clergyman. Nothing we know of Darwin at this time suggests that he was in any way exceptional or unconventional. There is no hint of the intellectual revolutionary he was later to become. In his own words: ". . . During the three years which I spent at Cambridge my time was wasted, as far as the academical studies were concerned, as completely as at Edinburgh and at school. . . ."[8]

There were, however, some pointers towards a career in the biological sciences. Throughout his youth he had been interested in natural history, as a boy he had collected insects, and later, at Cambridge, he made a serious hobby of entomology and made a few minor contributions to the field. Also, despite his confession to being a wastrel at Cambridge, there must have been a lot more to Darwin than a passion for riding and hunting. He befriended there some of the foremost intellectuals in the natural sciences, such as Professor

Henslow with whom he used to take long walks, earning for himself the title 'the man who walks with Henslow'. Through Henslow Darwin came in contact with Professor Sedgwick, one of the dominant figures in British geology at that time. Neither of these two men nor any of his other mentors were unconventional in any way. Nearly all were creationists and catastrophists who accepted a literal interpretation of Genesis. Darwin seems to have imbibed the conventional wisdom of his tutors.

Both before and even shortly after having joined the *Beagle*, Darwin was a Bible-quoting fundamentalist and a believer in the special creation and fixity of each species.

> . . . I did not then in the least doubt the strict and literal truth of every word in the Bible . . .[9]
>
> Whilst on board the *Beagle* I was quite orthodox, and I remember being heartily laughed at by several of the officers (though themselves orthodox) for quoting the Bible as a unanswerable authority on some point of morality.[10]

For Darwin the *Beagle* proved the turning point of his life, a liberating journey through time and space which freed him from the constraining influence of Genesis. Every voyage conjures up a vision of new horizons and emancipation, but there is something particularly evocative about the voyage of the *Beagle* to the remote and little known shores of South America. It is almost as if the elemental forces of nature, so apparent along those cold and stormy coasts of Patagonia and Tierra del Fuego, had conspired together to fragment the whole framework of biblical literalism in Darwin's mind, to blow his intellect clear of all the accumulated cobwebs of tradition and religious obscurantism. The *Beagle* is also symbolic of the much greater voyage which the whole of our culture subsequently made from the narrow fundamentalism of the Victorian era to the scepticism and uncertainty of the twentieth century. Darwin's experiences during those liberating five years became the experience of the world.

It was Darwin's geological observations on the *Beagle* which first sowed seeds of doubt in his mind as to the historicity of the Genesis account of creation. We know that Darwin had Lyell's *Principles of Geology* with him on the voyage and that it exerted a powerful influence on his thinking as the journey progressed. In Patagonia, while following the course of the Santa Cruz river, he considers the forces responsible for the carving out of a steep-sided canyon from the hard basalt rock and records in his journal:[11]

What power, then, had removed along a whole line of country, a solid mass of very hard rock, which had an average thickness of nearly three hundred feet, and a breadth varying from rather less than two miles to four miles? The river, though it has so little power in transporting even inconsiderable fragments, yet in the lapse of ages might produce by its gradual erosion an effect of which it is difficult to judge the amount. But in this case, independently of the insignificance of such an agency, good reasons can be assigned for believing that this valley was formerly occupied by an arm of the sea. . . . Geologists formerly would have brought into play, the violent action of some overwhelming debacle. But in this case such as supposition would have been quite inadmissible, because, the same step-like plains with existing seashells lying on their surface, which front the long line of the Patagonian coast, sweep up on each side of the valley of Santa Cruz. No possible action of any flood could thus have modelled the land, either within the valley or along the open coast; and by the formation of such step-like plains or terraces the valley itself has been hollowed out. Although we know that there are tides which run within the narrows of the Strait of Magellan at the rate of eight knots an hour, yet we must confess that it makes the head almost giddy to reflect on the number of years, century after century which the tides, unaided by a heavy surf, must have required to have corroded so vast an area and thickness of solid basaltic lava.

He must have had many similar experiences, as he explored the southern tip of South America, which would have made him sceptical of the theory of catastrophes and of the six thousand-year time scale and increasingly inclined to a uniformitarian view. In another section of his journal he reflects on the uplifting of the mighty Andes:[12]

Who can avoid wondering at the force which has upheaved these mountains, and even more so at the countless ages which it must have required, to have broken through, removed, and levelled whole masses of them?

Darwin's theory of coral reef formation, which he worked out while still on the west coast of South America, is further evidence that he had rejected the six thousand-year old Earth and was already, eighteen months before he returned to England, a convert to Lyell's theory. Coral grows very slowly and will only grow at a depth of about twenty feet below the surface of the sea, yet in some parts of the world vast reefs have been built up, some of them several hundred feet in height, reaching from the ocean floor to a few feet below the surface, consisting of myriads upon myriads of shells of minute and long dead coral polyps. Darwin noted that they were invariably found in regions where either land subsidence had occurred or where there was strong circumstantial evidence that it had probably occurred. He argued that if land subsided sufficiently slowly then the reef could grow upwards at the same rate and over a course of an unimaginable span of time an immense reef, perhaps several hundred feet in depth, could be gradually built up. Such a process would require that geological change, in this case land subsidence, occurred very gradually and over immense spans of time, and these two axioms, gradualism and an immense time span, were the two fundamental pillars upon which the whole uniformitarian thesis of geology was based.

The twin concepts of gradualism and immense time are also crucial to the idea of biological evolution and, as many biologists later acknowledged, geological uniformitarianism, more than anything else, eased the way for their acceptance of evolution. Whether Darwin himself made the transition while on board the *Beagle* is difficult to assess from his own writings. Precisely when he came to believe in evolution, whether it was a gradual dawning, or a sudden realization, we will probably never know. What is certain, however, is that the biological observations he made on the voyage, particularly those relating to geographical variation, played a crucial role in the development of his evolutionary thinking.

The one aspect of geographical variation which more than any other seemed to challenge the concept of the fixity of species was the fact that different geographical regions were often populated by quite distinct, yet closely related, species. Only a few months after leaving England, the *Beagle* was bringing Darwin face to face with many striking examples of this phenomenon. After making a short expedition into the Brazilian jungle he recorded in his journal:[13]

I never returned from these excursions empty-handed. This day I found a specimen of a curious fungus, call Hymenophallus. Most people know the English Phallus, which in autumn taints the air with its odious smell: this, however, as the entomologist is aware, is to some of our beetles a delightful fragrance. So was it here; for a Stronglylus, attracted by the odour, alighted on the fungus as I carried it in my hand. We here see in two distant countries a similar relation between plants and insects of the same families, though the species of both are different.

Such observations present an obvious challenge to the doctrine of the fixity of species, and this must have been perceived by Darwin. A number of questions automatically arise: were such clearly related species really created separately in Europe and South America as a rigid application of the fixity of species would imply, or had they gradually diverged by natural processes from a common stock as they migrated to their present geographical locations? If they had diverged, would there have been sufficient time in six thousand years for the degree of change to occur?

Darwin must have had serious doubts as to the validity of the fixity of species before he reached the Galapagos Islands but it seems as if the observations he made there finally provided him with irrefutable evidence that the species was not an immutable entity; and this was his moment of truth from which there could be no turning back. Ahead lay the intellectual path to evolution and ultimately *The Origin of Species*. Darwin, unfortunately, left no record of his exact state of mind over the question of evolution while he was on the Galapagos Islands but we know that in retrospect he judged the experiences he had there to be the "origin of all my views".[14]

The Galapagos Archipelago consists of thirteen small volcanic islands situated on the equator about six hundred miles west of the coast of South America. The largest is only seventy miles long and only twenty miles wide at its broadest point, while some of the smaller islands are no more than a few square miles in area. The majority of the islands are less than sixty miles apart. The Archipelago is not a particularly attractive backdrop for the enactment of a decisive intellectual drama. As Darwin records his first impressions of one of the main islands of the group:[15]

Nothing could be less inviting than the first appearance. A broken field of black basaltic lava, thrown into the most rugged waves and crossed by great fissures, is everywhere covered by stunted, sunburnt brushwood, which shows little signs of life. The dry and parched surface, being heated by the noonday sun, gave to the air a close and sultry feeling, like that from a stove. We fancied even that the bushes smelt unpleasantly. Although I diligently tried to collect as many plants as possible, I succeeded in getting very few, and such wretched-looking little weeds would have better become an arctic than an equatorial flora.

But if the physical appearance of the remote islands was uninviting, the natural history of the Archipelago was, in Darwin's own words, 'eminently curious and well deserves attention', as the islands were populated by a remarkable number of unique and unusual plant and animal species. As Darwin recorded in his journal, these included at least one hundred species of flowering plants, dozens of insect species, and nearly thirty species of bird. Also found only on the Archipelago were a unique species of giant tortoise and two closely related species of lizards, one terrestrial and the other, the remarkable marine iguana, uniquely specialized for feeding on seaweed, possessing partially webbed feet and the capacity to remain submerged for prolonged periods of time. One of the most intriguing aspects of the fauna of the Galapagos Islands was the way many of the organisms such as the tortoises, the iguanas, the mocking thrushes and many of the plants varied from island to island; and in some cases this variation was so marked that individual island forms had every appearance of being quite distinct species.

It was the existence of so many distinct, yet intimately related, species scattered on the islands of the Archipelago which planted the idea of organic evolution in Darwin's mind. As he noted in his journal:[16]

> The distribution of the tenants of this archipelago would not be nearly so wonderful, if, for instance, one island had a mocking-thrush, and a second island some other quite distinct genus; – if one island had its genus of lizard, and a second island another distinct genus, or none whatever – or if the different islands were inhabited, not by representative species of the same genera of plants, but by totally different genera. . . . But it is the circumstance, that several of the islands possess their own species of the tortoise, mocking-thrush, finches, and numerous plants, these species having the same general habits, occupying analogous situations, and obviously filling the same place in the natural economy of this archipelago, that strikes me with wonder.

The possibility that the closely related species on the different islands had descended with modification, or evolved, from a common

ancestral species which originally inhabited the islands, was very
difficult to resist. Was it really possible, as a strict application of the
doctrine of the fixity of species implied, that individual species had
been specifically created for tiny islands, some of which were in
Darwin's own words hardly more than 'points of rock'? As he wrote
in his journal:[17]

> . . . one is astonished at the amount of creative force, if such an
> expression may be used, displayed on these small, barren, and rocky
> islands; and still more so, as its diverse yet analogous action on points
> so near each other.

In his ornithological notes composed while on board the *Beagle*,
perhaps even shortly after he left the Galapagos Islands, Darwin
gives us a glimpse into his own mind, where the idea of evolution was
clearly beginning to dawn. Commenting on the inter-island variation
he had just witnessed he remarked:[18]

> . . . the zoology of Archipelagos . . . will be well worth examining; for
> such facts would undermine the stability of Species.

Of all the animals unique to these remote islands perhaps none are
more famous than a group of small land birds now known to the
world as 'Darwin's Finches'.

Altogether there are fourteen different species of finches on the
Galapagos Islands which differ so greatly in size, plumage, beak
morphology and behaviour that were they to visit an average suburban
garden they would be classed unhesitatingly as distinct species. The
largest finch is about the size of a blackbird while the smallest is close
to a sparrow. Each has a distinctive plumage. In some it is almost
completely black while in others it is light brown. The shape of the
beak varies markedly between the species: some have small finch-like
beaks, others parrot-like beaks, another group have slender warbler-
like beaks some of which are specially decurved for flower probing.
One species even has a straight wood-boring beak. The variation in
beak morphology reflects fundamental differences in feeding habits
and general behaviour. Some species, those with heavily built finch
or parrot-like beaks, the ground finches, are seed and cactus eaters,
and spend most of their time hopping about on the ground. The
species with long slender beaks, the tree finches, are largely insec-

Figure 1.1: The Galapagos Finches. *Some of the different species of finches on the Galapagos Islands, all of which originated from a common ancestor.* (from Lack)[19]

tivorous and spend much of their time tit-like among the branches of
the trees. One species, the woodpecker finch with a straight wood
boring beak, climbs woodpecker-like vertically up tree trunks and
has invented the remarkable technique of inserting a cactus spine
into crevices in search of insects. The warbler finch has a long slender
beak and not only exhibits the quick flitting movements of a warbler
as it darts among the branches searching for insects, but also like a
warbler repeatedly flicks the wings partly open when hopping among
the bushes. Although differing in coloration, beak morphology, feeding habits and size, the fourteen species of Galapagos finches are
undoubtedly closely related. All the species, for example, exhibit
exactly the same display and song pattern and all belong to the same
subfamily of finches.

So here on this isolated archipelago was a unique set of distinct
finches so closely related that some of them could be arranged into an
almost perfect morphological sequence in terms of beak morphology,
size and plumage. The idea that they were all related by common
descent from an original ancestral species, in other words that new
species had arisen from pre-existing species in nature and that,
therefore, species were not the fixed immutable entities most biologists
supposed, seemed irresistible. As Darwin wrote:[20]

> Seeing this gradation and diversity of structure in one small, intimately
> related group of birds, one might really fancy that from an original
> paucity of birds in this archipelago, one species had been taken and
> modified for different ends.

In addition to the remarkable interspecies variation within the
Archipelago, another aspect of the natural history of the islands
which lent further circumstantial support to the concept of evolution
and mitigated greatly against the doctrine of the fixity of species was
the very suggestive observation that despite the uniqueness of the
fauna of the Galapagos most of the species there were obviously, if
distantly, related to sister species on the nearest continental land
mass, the South American mainland some six hundred miles to the
east. Commenting on the relationship Darwin wrote in his journal:[21]

> If this character were owing merely to immigrants from America,
> there would be little remarkable in it; but we see that a vast majority of
> all the land animals, and that more than half of the flowering plants,

are aboriginal productions. It was most striking to be surrounded by new birds, new reptiles, new insects, new plants, and yet by innumerable triffling details of structure, and even by the tones of voice and plumage of the birds, to have the temperate plains of Patagonia, or the hot dry deserts of Northern Chile, vividly brought before my eyes.

Explaining why the fauna on this isolated archipelago should bear the unmistakable impression of South America in terms of creationism and the fixity of species seemed to lead to highly implausible conclusions. Why, on creationist reasoning, should the fauna of the Galapagos Islands resemble the fauna of South America and not, for example, the fauna of the Cape Verde Islands which are far closer in climate, geology, and general characteristics? In Darwin's own words:[22]

Why, on these small points of land, which within a late geological period must have been covered by the ocean, which are formed of basaltic lava, and therefore differ in geological character from the American continent, and which are placed under a peculiar climate, – why were their aboriginal inhabitants, associated, I may add, in different proportions both in kind and number from those on the continent, and therefore acting on each other in a different manner – why were they created on American types of organization? It is probable that the islands of the Cape de Verd group resemble, in all their physical conditions, far more closely the Galapagos Islands than these latter physically resemble the coast of America; yet the aboriginal inhabitants of the two groups are totally unlike; those of the Cape de Verd Islands bearing the impress of Africa, as the inhabitants of the Galapagos Archipelago are stamped with that of America.

This phenomenon was not restricted to the Galapagos Islands. To any well travelled naturalist it is immediately apparent that in different continents similar environments are generally occupied by quite different unrelated species and that adjacent geographical regions within any one great continental area are generally populated by different, yet basically related, forms. Why had God not created the same species for the same environments even if these environments did occur in widely separated geographical regions? Perhaps creation had proceeded according to some geographical rule which demanded that only closely related species be created within any one great region of the earth. Or had the curious pattern of geographical variation resulted from some sort of directed migration following the deluge? Such questions were bound to have occurred to Darwin while still on board the *Beagle*, and they must inevitably have had the effect of rendering the biblical framework increasingly obsolete in his mind.

Once it was accepted that new species could arise in nature by descent from pre-existing species, as seemed to have occurred within the Galapagos Archipelago itself, many of the facts of geographical variation could be readily explained. The close relationship between the fauna of the Galapagos and South America, for example, could then be easily accounted for by envisaging, firstly, a number of original chance colonizations of the islands from the South American mainland, and secondly, their subsequent evolutionary diversification into various new species as they spread gradually throughout the Archipelago. In many such instances Darwin must have found the evolutionary explanation far more plausible than its creationist rival. Altogether, many of the facts of geographical variation were very difficult to reconcile with the doctrine of the fixity of species and later, in a much quoted letter to Joseph Hooker on 11th January, 1844, Darwin made his famous confession:[23]

> I was so struck with the distribution of the Galapagos organisms, &c. &c., and with the character of the American fossil mammifers &c. &c., that I determined to collect blindly every sort of fact, which could bear any way on what are species. . . . At last gleams of light have come, and I am almost convinced (quite contrary to the opinion I started with) that species are not (it is like confessing a murder) immutable.

He was not the only Victorian naturalist whose faith in the fixity of species was shaken by travel and, particularly, by contact with the facts of geographical variation in isolated regions. Lyell, whose book had such an influence on Darwin's geological thinking, but who resisted the idea of organic evolution for many years, first felt the impact of Darwin's argument when he too had been exposed to the phenomenon of geographical variation on the Canary Islands. Similarly, Alfred Russel Wallace, who subsequently read with Darwin their famous joint paper to the Linnean Society in 1858 proposing the theory of evolution by natural selection, first became an evolutionist when he became acquainted with the facts of geographical variation in Malaya and in the Indonesian Islands.

The *Beagle* revealed to Darwin a new world, one that bore no trace of the supernatural drama that Genesis implied, and one which seemed impossible to reconcile with the miraculous biblical framework he himself had accepted when he left England. All the new evidence seemed to point to an immensely long geological past, and nowhere could he see evidence of supernatural catastrophes or interventions interrupting the course of nature. The doctrine of the fixity of species was contradicted by the sorts of observations he had made on the Galapagos Islands which suggested strongly that species did in fact change under the agency of entirely natural processes.

Although nothing that Darwin had witnessed on the *Beagle* implied that evolution on a grand scale had occurred, that the major divisions of nature had been crossed by an evolutionary process, the old typological discontinuous view of nature seemed far less credible. This was not only because the species barrier, one of the supposedly fundamental divisions of nature, had apparently been breached in places like the Galapagos Islands, but also because in Darwin's mind and in the minds of many nineteenth-century biologists, typology was closely associated with the whole supernatural biblical framework with its emphasis on a recent Earth, on the miraculous and special creationism, a framework which was frankly non-scientific and irreconcilable with the fundamental aim of science to reduce wherever possible all phenomena to purely natural explanations.

NOTES

1. Coleman, W. (1964) *Georges Cuvier: Zoologist*, Harvard University Press, Cambridge, Mass. pp 172–73.
2. Mayr, E. (1972) "The Nature of the Darwinian Revolution", *Science*, 166; 981–89, see pp 983–84.
3. The Zoological Journal (1824) 1: 7.
4. Agassiz, L. (1857) *Essay on Classification*, new ed (1962) with introduction by E. Lurie, Harvard University Press, Cambridge, Mass, p 137.
5. Coleman, op cit, pp 170–71.
6. Cuvier, G. (1812) *Rescherches sur les ossemens fossiles de quadrupeds*, vol 1, pp 8–9, cited in and translated by C.C. Gillispie (1959) *Genesis and Geology*, Harper and Row, New York, p 100.
7. Chalmers, T. (1814) "Remarks on Cuvier's Theory of the Earth" in Miscellanies: *Embracing Reviews, Essays and Addresses* (1848) 4 vols, Robert Carter, New York, vol 1, p 191.
8. Darwin, F. (1888) ed, *The Life and Letters of Charles Darwin*, 3 vols, John Murray, London, vol 1, p 46.

9. ibid, vol 1, p 45.
10. ibid, vol 1, p 307.
11. Darwin, C. (1845) The voyage of the Beagle, new ed (1959) by J.M. Dent, London, p 172.
12. ibid, p 245.
13. ibid, p 31.
14. Darwin, F., op cit, vol 1, p 276.
15. Darwin, C., op cit, p 359.
16. ibid, p 382.
17. ibid, p 383.
18. Barlow, N., (1963) "Darwin's Ornithological Notes", *Bull. of the British Museum (Natural History)*, Historical Series, vol 2, p 203–78, see p 262.
19. Lack, D. (1974) *Darwin's Finches*, Cambridge University Press, Cambridge, from Figure 3 and Plates III, IV and V and frontispiece.
20. Darwin, op cit, p 365.
21. ibid, p 378.
22. ibid, p 378.
23. Darwin, F., op cit, vol 2, p 23.

CHAPTER 2
The Theory of Evolution

> There is grandeur in this view of life, with its several powers, having
> been originally breathed by the Creator into a few forms or into one;
> and that, whilst this planet has gone cycling on according to the fixed
> law of gravity, from so simple a beginning endless forms most beautiful
> and most wonderful have been, and are being evolved.

Revolutionary though it was in Victorian England, there was nothing
fundamentally novel about the central concept of Darwinian theory.
The core idea of the *Origin*, the idea that living things have originated
gradually as a result of the interplay of chance and selection, has a
long pedigree. It can be traced back from the views of current
advocates of Darwinian orthodoxy such as Huxley, Mayr and Simpson
to Darwin and from Darwin via Hume in the eighteenth century
right back to the materialistic philosophers of classical times. The
idea is clearly expressed in the philosophies of Democritus and
Epicurus and by many of the even older Ionian nature philosphers
such as Empedocles.

Even some primitive mythologies express the idea that life in all its
diverse manifestations is not the creation of the gods but a purely
natural phenomenon being the result of normal flux of the world.
The ancient Norse, for example, held that the first living beings, the
giant Ymir and the primordial cow Audumla, were formed gradually
from the ice melted by the action of a warm wind which blew from a
southern land Muspellsheim, the land of fire.

The majority of the old pre-Socratic philosophers were strikingly
materialistic in their interpretation of nature. To them life was a
natural phenomenon, the result of processes no less natural than
those which moulded the forms of rocks or rivers, no less inevitable
than the turn of the tides, the phases of the moon. Life was for them
part of a continuum with the soil and sea.

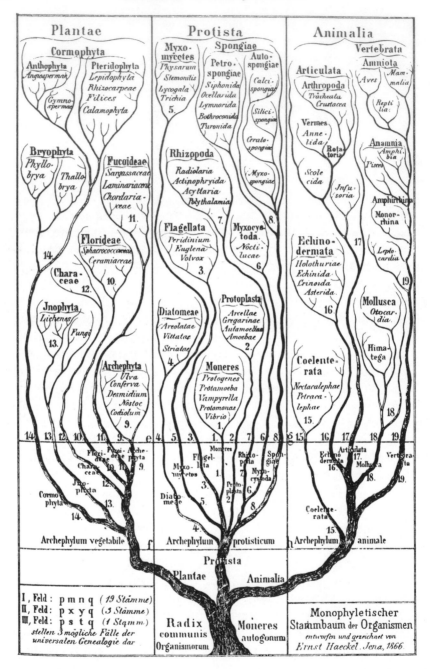

Figure 2.1: The First Evolutionary Tree of Life. *Drawn up in 1866 by Ernst Haeckel.*[1]

According to Anaximander of Miletus (550 BC), one of the earliest of the nature philosophers, life was first generated by material processes from sea slime, an idea remarkably reminiscent of modern theories of the origin of life and the pre-biotic soup. In this idea he was probably influenced by the Homeric poems, where man was considered a mixture of water and earth, as well as by the empirical observation that the body is composed of solid and fluid elements. The primeval aquatic creatures which emerged from the marine slime were supposedly possessed of a bristly integument which they cast off like insect larvae as they evolved into terrestrial forms. Anaximander's evolutionary philosophy was one of the first rational attempts to give a naturalistic explanation of the origin and diversity of life. His philosophy expresses both the idea of a purely natural origin of life and the idea of descent with modification. However, Anaximander's theory was still incomplete in one important aspect: it failed to deal with the problem of organic design.

But already by late pre-Socratic times the materialists were able to avoid teleological conclusions by proposing a naturalistic explanation for adaptive design. The mechanism they hit on was essentially a primitive form of natural selection, and selectionist explanations for the teleonomy of living things have been basic to all materialistic theories since.

Empedocles (450 BC) was one of the first of the materialists to realize that the phenomenon of adaptive complexity required a specific explanation within the framework of a naturalistic scheme. His selectionist theory to account for the design of organisms preceded Darwin by two thousand years. Empedocles supposed that as a result of the continual flux of matter all sorts of fantastic shapes and objects were continually generated by chance interaction of elements. Heads, for instance, without neck and trunk, eyes without a face, arms without shoulders would occasionally be joined together often creating monstrosities, double-headed and double-breasted beings, human heads with the bodies of bulls, bodies of bulls with human heads and so forth. Theodore Gomperz comments:[2]

These grotesque shapes disappeared as quickly as the original separate limbs, and only such combinations as exhibited an inner harmony evinced themselves as fit for life, maintained a permanent place, and finally multiplied by procreation. It is impossible not to be reminded here of the Darwinian survival of the fittest. There is nothing to

prevent and everything to favour the belief that we are confronted
with an attempt, as crude as it could be, but not yet entirely unworthy
of respect, to explain in a natural way the problem of design in the
organic world.

The pre-Socratic materialistic concept of life reached its most
perfect expression in the philosophies of the so-called Atomists such
as Democritus and Epicurus; already by the fifth century BC the
gods had been declared unnecessary and the two basic concepts
which underlie modern evolutionary thought had been clearly formulated. Firstly, there was the idea of the continuity of nature, that
the living kingdom is composed of an ever-changing collection of
organismic forms all related by descent from a primordial progenitor;
and secondly, there was the idea of the selection of random changes as
the primary generative process responsible for the creation and
adaptive shaping of the ever-changing spectrum of life.

The philosophy of the pre-Socratic materialists laid the foundation
upon which all subsequent naturalistic speculation regarding the
origin and design of life was based. For example, Hume's explanation
to account for the design of living things, which he gives in his
famous dialogue concerning natural religion, is basically the same.
He proposed that the world was composed of a finite number of
particles which were in perpetual random motion. In unlimited time
the particles enter into every combination possible. Occasionally
they enter into stable conformations which tend to persist.[3]

> A finite number of particles is only susceptible of finite transpositions,
> and it must happen in an eternal duration that every possible order or
> position must be tried an infinite number of times . . . the continual
> motion of matter, therefore, in less than infinite transpositions must
> produce this economy or order and by its very nature that order, when
> once established supports itself for the many ages.

So, according to Hume, the random juggling of matter must eventually produce ordered forms adapted to their environment and possessing an intrinsic coherence in their components which gives the
appearance of design.

In the century before Darwin most zoologists who had toyed with
the idea of evolution had departed from this materialistic tradition.
Their theories were mostly speculative and invariably postulated

"non-material, inner forces" or "vital drives" of a basically mysterious nature which lay deep within organisms and which were presumed to drive evolution along inexorably progressive paths to ever more complex and perfect ends. There was, of course, not a scrap of evidence in support of these occult and vitalistic theories, one of the best known of which is the hypothesis of the Frenchman Lamarck, whom Darwin acknowledged in the introduction to the *Origin*.

This is not the place for a lengthy and detailed discussion of Lamarck's theory but since many today still conceive of evolution in Lamarkian terms, and because there exists a degree of confusion between Lamarck's views and Darwin's theory of natural selection, it is worth describing briefly. Lamarck proposed that improvements acquired by an individual during its lifetime could be passed on to its offspring and so gradually, as each successive generation strove to improve its characteristics, adaptive perfection was achieved. Hence, according to Lamarck, the long neck of the giraffe evolved because the original ancestors of the modern giraffe, endowed with necks no longer than a cow, in attempting to reach leaves high above the ground managed to stretch their necks to make them longer. This acquired characteristic was then passed on to their offspring who were born with slightly lengthened necks. And so the process continued until after many generations of striving by the animals to reach ever higher leaves the long neck of the giraffe evolved, perfectly adapted for grazing on the leaves of the tallest trees. Thus the 'dream' of the organism eventually found concrete expression. Lamarck's theory, like all the other vitalistic evolution theories, necessitated the existence of some sort of mysterious intelligent feed-back device in every living organism which could directly influence the genetic make-up of its offspring in a particular and intelligent way so that its adaptations could be purposefully changed and improved. Thus evolutionary change was directed according to the requirement of the organisms.

None of these vitalistic theories of evolution held any appeal for Darwin. He was from the beginning far too scientific in outlook and was diametrically opposed to any notion that there were deep seated forces of an occult, unknown nature which perfected adaptations and guided evolution towards particular ends and goals.

When the *Beagle* docked in Falmouth in October 1836, although he had accumulated most of the evidence he needed to build his evolutionary view of nature, Darwin was still puzzled as to what mechanism might make evolutionary change come about. We know from his own writings that he was already well acquainted with the way in which artificial selection had been applied by man to the creation of new races of domestic animals and plants but he had not grasped fully the potentiality of selection as an agency for change in nature. It was not until he read Malthus in October 1838, two years after he had left the *Beagle,* that he fully grasped the significance of the fact that in each generation "more individuals are produced than can possibly survive" – which implied that all living things were subject to an intense struggle for existence and were under a profound and continuing selection pressure so that only the very fittest or best survived. The Malthus insight seems to have been one of those classic flashes of scientific inspiration. In his autobiography he describes how suddenly he "at last got a theory by which to work".[4]

> In October 1838 . . . I happened to read for amusement Malthus on *Population*, and being well prepared to appreciate the struggle for exist-ence which everywhere goes on from long-continued observation of the habits of animals and plants, it at once struck me that under these circumstances favourable variations would tend to be preserved, and unfavourable ones to be destroyed. The result of this would be the formation of new species. Here, then, I had at last got a theory by which to work; but I was so anxious to avoid prejudice, that I determined not for some time to write even the briefest sketch of it. In June 1842 I first allowed myself the satisfaction of writing a very brief abstract of my theory in pencil in thirty-five pages . . .

The concept of evolution by natural selection was elegant and beautifully simple. It avoided completely the necessity to propose the "inner drives" which were so characteristic of pre-Darwinian theories. In essence, the mechanism depended on only three premises each of which were practically self evident: that organisms varied; that these variations could be inherited; and that all organisms were subject to an intense struggle for existence which was bound to favour the preservation by natural selection of beneficial variations. Given variations, given that they could be inherited, and given natural selec-tion, then evolution and adaptive change were inevitable. Darwin had an evolutionary theory that was entirely materialistic and mechanistic.

The evolution of the long neck of the giraffe could now be explained without recourse to mysterious "inner forces". It could now be pro-posed that purely by chance some individuals were born with for-

tuitously slightly longer necks, and that this conferred upon them a
selective advantage enabling them to reach higher branches in times
of famine and drought, which greatly improved their chances of surviving and leaving offspring similarly endowed with longer necks.
Such a process repeated over many generations would inevitably lead
to the long neck of the modern giraffe.

It is important at this stage to be clear about Darwin's view of
variation, the raw material of evolution. Although the mechanism of
heredity was not understood in Darwin's day, it was self evident that
individual organisms were not identical but varied in a number of
different ways: some individuals were slightly taller than others,
some had slightly different colours and so on. Darwin believed, and
we now know that he was correct, that the mechanism responsible for
these genetic variations was entirely blind to the adaptive needs and
requirements of the organism. If a beneficial variation occurred
which conferred upon an organism some slight adaptive advantage or
improvement this was entirely fortuitous. In other words the changes
were undirected and as likely to be detrimental or neutral to the
organism's survival as beneficial. The purely random nature of the
mutational input or the direction of variation served to differentiate
Darwin's theory from all the other vitalistic evolutionary theories
such as Lamarck's, for in all these pre-Darwinian theories variations
are not random but rather directed, adaptive and purposeful. Ulti
mately, Darwin's theory implied that all evolution had come about
by the interactions of two basic processes, random mutation and
natural selection, and it meant that the ends arrived at were entirely
the result of a succession of chance events.

Evolution by natural selection is therefore, in essence, strictly
analogous to problem solving by trial and error, and it leads to the
immense claim that all the design in the biosphere is ultimately the
fortuitous outcome of an entirely blind random process – a giant
lottery. Thus Darwin was proposing, as Jacques Monod has put it:[5]

> . . . that chance *alone* is at the source of every innovation, of all creation
> in the biosphere. Pure chance, absolutely free but blind, at the very
> root of the stupendous edifice of evolution . . .

It was a revolutionary claim. Where once design had been the result
of God's creation, it was now put down to chance.

Armed with a mechanism and with the evidence he had gained on the *Beagle*, Darwin was ready to construct his great synthesis. Only a few months after his "Malthus insight" and shortly before his marriage in January 1839, his views had so matured that his notebooks at the time contained all the core ideas which twenty years later he announced to the world in *The Origin of Species*.

The Origin of Species has been referred to as "one of the most important books ever written"[6] and "a book that shook the world";[7] seldom have such superlatives been more appropriate. Even after the lapse of a century it is still, in the words of one modern authority on evolution theory,[8] "the best and most interesting general treatment of evolutionary biology ever written." The lack of a logical structure in the Origin makes it a difficult book to summarize but the main arguments are clear enough.

In his book Darwin is actually presenting two related but quite distinct theories. The first, which has sometimes been called the "special theory", is relatively conservative and restricted in scope and merely proposes that new races and species arise in nature by the agency of natural selection, thus the complete title of his book: *The Origin of Species by Means of Natural Selection or the Preservation of Favoured Races in the Struggle for Life*. The second theory, which is often called the "general theory", is far more radical. It makes the claim that the "special theory" applies universally; and hence that the appearance of all the manifold diversity of life on Earth can be explained by a simple extrapolation of the processes which bring about relatively trivial changes such as those seen on the Galapagos Islands. This "general theory" is what most people think of when they refer to evolution theory.

The first five chapters deal mainly with evidence for the special theory and microevolutionary phenomena. One of the primary goals Darwin was aiming at in these chapters was to demolish the concept of the immutability of species, and no one who has read the *Origin* can deny the skill and force with which he marshals the evidence and presents his arguments. Chapter One, "Variation under Domestication", discusses the power of selection in the hands of man as evidenced by the tremendous degree of variation that has been produced by the selective breeding of different races of domestic animals. In the case of the various breeds of pigeon, Darwin makes the very telling point that many of the varieties, although all descended from the common rock pigeon, are so different that most ornithologists would have to class them as separate species![9]

The variation of domestic animals provided Darwin not only with evidence of the power of selection but also with irrefutable evidence that organisms could indeed undergo a considerable degree of evolutionary change. The fact that the differences between domestic breeds appeared to be as great as that between species in nature obviously tended to undermine the concept of the immutability of the species. If domestic breeds had descended from a common ancestral type, might not some related species have likewise descended with modification from an ancestral type?

In Chapter Two, "Variation in Nature", Darwin shows how arbitrary are the criteria used by zoologists in designating a group of individuals as a species and how the same criteria could just as easily be applied to varieties in some instances and groups of species in another.

In the following two chapters he describes the struggle for existence and its consequence, natural selection, to which all living things are inevitably subject. He concedes that selection in nature would be a far less efficient cause of change than selection in the hands of man, but argues that given a sufficient period of time there is no reason to suppose that it might not generate as great, if not far greater, a degree of change.

As we have seen (see Chapter One), Darwin saw the phenomenon of geographical variation, especially as witnessed on isolated oceanic islands like the Galapagos, as providing powerful support for the idea that new species had evolved from pre-existing species. It was the facts of geographical variation more than anything else which he considered to be the "origin of all my views", and interestingly, as if to emphasise the point, the very first paragraph of the introduction to the *Origin* refers to the same phenomenon:[10]

> When on board H.M.S. *Beagle*, as naturalist, I was much struck with certain facts in the distribution of the organic beings inhabiting South America. . . . These facts, as will be seen in the latter chapters of the volume, seemed to throw some light on the origin of species . . .

Darwin dealt with the facts of geographical variation systematically in Chapters Twelve and Thirteen of the *Origin*. Here, too, he again returns to his experiences on the *Beagle* and to the distribution of species in South America and the Galapagos Islands. It is obvious from the self-assured style of these chapters that he viewed the phenomenon of geographical variation as providing one of his most convincing arguments against the immutability of species and the creationist view of nature. Here was what appeared to be compelling evidence that, at least in some cases, the origin of species was not a supernatural event but the result of a perfectly natural process of descent with modification; and, although Darwin was usually guarded with regard to his evolutionary claims, as far as the evidence of geographical variation was concerned he was prepared to be uncompromising:[11]

> Facts such as these admit of no sort of explanation on the ordinary view of independent creation . . . such facts as the striking relationship between the inhabitants of islands and those of the nearest mainland – the still closer relationship of the distinct inhabitants of the islands in the same archipelago – are inexplicable on the ordinary view of the independent creation of each species, but are explicable if we admit colonisation from the nearest or readiest source, together with the subsequent adaptation of the colonists to their new homes.[12]

Although all Darwin's evidence, even the evidence of geographical variation, was in the last analysis entirely circumstantial, nevertheless, the arguments and observations he assembled in the first five chapters, as well as in Chapters Twelve and Thirteen, enabled him to build a very convincing case for his special theory – that speciation, the origin of new species from pre-existing species, can, and does, occur in nature as a result of perfectly natural processes in which natural selection plays a key role.

If the *Origin* had dealt only with the evolution of new species it would never have had its revolutionary impact. It was only because it went much further to argue the general thesis that the same simple natural processes which had brought about the diversity of the Galapagos finches had ultimately brought forth all the diversity of life on earth and all the adaptive design of living things that the book proved such a watershed in western thought. Much of the *Origin*, especially the later chapters, dealt not with the special theory which gave the book its title, but with a defence of its general application.

One of the key arguments Darwin advances, and one to which he returns at least implicitly in many places in the *Origin*, is that once it is conceded that organisms are inherently capable of a considerable degree of evolutionary change, then might they not, especially if a great length of time is allowed, be potentially capable of undergoing practically unlimited change sufficient even to bridge some of the seemingly most fundamental divisions of nature?

The convincing arguments and evidence Darwin had assembled to show that the species barrier was not the unbridgeable discontinuity that the typologists maintained had enormous psychological impact, because the doctrine of the fixity of species was the cornerstone of the whole typological world view. As Mayr points out:[13]

> Darwin's choice of title for his great evolutionary classic, *On the Origin of Species*, was no accident. The origin of new "varieties" within species had been taken for granted since the time of the Greeks. . . . The species remained the great fortress of stability and this stability was the crux of the antievolutionist argument.

The threat to typology if the doctrine of the fixity of species was abandoned was obvious to Darwin:[14]

> Several eminent naturalists have of late published their belief that a multitude of reputed species in each genus are not real species; but that other species are real, that is, have been independently created. This seems to me a strange conclusion to arrive at. They admit that a multitude of forms, which till lately they themselves thought were special creations, and which are still thus looked at by the majority of naturalists, and which consequently have all the external characteristic features of true species – they admit that these have been produced by variation, but they refuse to extend the same view to other and slightly different forms.

With the notion of a static world order threatened in this way, it was far more difficult to accept the typological insistence that the divisions in nature were unbridgeable, that biological variation was always conservative and change always circumscribed by the boundaries of the type. Now it could be considered limitless and radical and therefore potentially capable of generating any degree of evolutionary change or innovation.

Admittedly, the length of time necessary for evolution by natural selection would have to be very great. As Darwin concedes in the *Origin*, the greater the length of time, the "better chance of beneficial variations arising and of their being selected accumulated, and fixed."[15] However, the revelation of just how vast an amount of time had elapsed during the history of life on earth was one of the major

discoveries of nineteenth-century geology. Attempting to compre-
hend this immensity impresses the mind, as Darwin put it, "almost
in the same manner as does the vain endeavour to grapple with the
idea of eternity".[16] The undeniable capacity of organisms to undergo
at least a degree of change, taken in conjunction with the vast time
available, seemed to Darwin to greatly enhance the plausibility of his
macroevolutionary claims. In an eternity, any degree of change might
occur. Organisms obviously underwent changes which could be inherited. If a pouter and a fantail pigeon could all be derived from a
rock dove then why, given very much greater periods of time, could a
horse and an octopus not have been similarly derived from an amoeba?
As he wrote in the *Origin*:[17]

> I can see no limit to the amount of change, to the beauty and complexity
> of the co-adaptations between all organic beings, one with another and
> with their physical conditions of life, which may have been affected in
> the long course of time through nature's power of selection, that is by
> the survival of the fittest.

In addition, Darwin was able to allude to a great deal of evidence
drawn from the fields of comparative anatomy and paleontology which,
he argued, were highly suggestive of the reality of macroevolution
and appeared to confirm the validity of the extrapolation. One of the
most important discoveries which arose from the study of comparative
anatomy was the realization that the resemblances between living
organisms are of two quite different sorts. On the one hand, there is
the sort of resemblance where a fundamentally dissimilar structure
has been modified or adapted to similar ends. This is known as analogous resemblance and is seen, for example, in the similarity between
the flipper of a whale and the fin of a fish, between the forelimbs of a
mole and those of a mole cricket. The other sort of resemblance is
termed homologous and occurs where a fundamentally similar organ
or structure is modified to serve quite dissimilar ends. A good example
of homologous resemblance is the similarity in the basic design of the
forelimbs of terrestrial vertebrates.

Darwin felt that homology was highly suggestive of the reality of
macroevolution:[18]

> What can be more curious than that the hand of man, formed for
> grasping, that of a mole for digging, the leg of the horse, the paddle of

the porpoise, and the wing of the bat, should all be constructed on the same pattern, and should include similar bones, in the same relative positions?

We may call this conformity to type, without getting much nearer to an explanation of the phenomenon . . . but is it not powerfully suggestive of true relationships, of inheritance from a common ancestor?

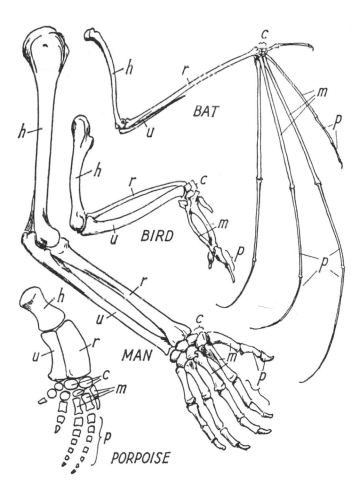

Figure 2.2: The Phenomenon of Homology. *Illustrated by the resemblance in the structure of the forelimbs of a number of vertebrate species.*

h = humerus; r = radius; u = ulna; c = carpals; m = metacarpals; and p = phalanges (from Hardy)[19]

One aspect of homology was especially difficult to account for by a theory of special creation and that was the existence of rudimentary organs.[20]

> It would be impossible to name one of the higher animals in which some part or other is not in a rudimentary condition. In the mammalia, for instance, the males possess rudimentary mammae; in snakes one lobe of the lungs is rudimentary; in birds the "bastard-wing" may safely be considered as a rudimentary digit, and in some species the whole wing is so far rudimentary that it cannot be used for flight. What can be more curious than the presence of teeth in foetal whales, which when grown up have not a tooth in their heads; or the teeth, which never cut through the gums, in the upper jaws of unborn calves?

Comparative anatomy not only revealed an underlying unity in the design of each group of organisms, it also revealed that the different species within each group possessed unique characteristics, and that the distribution of these characteristics conformed to a highly ordered pattern which permitted the species to be classified into a hierarchy of increasingly inclusive classes.

The diagram below shows a classification scheme for a group of related mammalian species. They can be grouped on shared similarities into different categories – species, genera families, orders – and these categories can be arranged into a hierarchy of ascending rank, smaller subgroups being ordered within successively larger subgroups.*

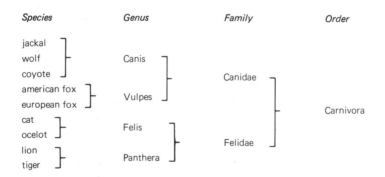

*Since classical times it has been known that living organisms could be classified according to their similarities of form and structure into several major divisions known as *phyla* and that these could be successively subdivided into smaller and

The pre-Darwinian biologists believed that all of the categories represented an ideal plan or type which had been conceived in the mind of God. All the members of a particular category represented expressions of a basic eternal theme which was a fixed and unchanging part of God's plan for creation. Thus all mammals were variations on the mammalian theme, all cats variations on the cat theme and so on.

Darwin argued that evolution provided a radically different but highly satisfying explanation of the hierarchic order of nature:[21]

> . . . if I do not greatly deceive myself, on the view that the Natural System if founded on descent with modification; – that the characters which naturalists consider as showing true affinity between any two or more species, are those which have been inherited from a common parent, all true classification being genealogical; – that community of descent is the hidden bond which naturalists have been unconsciously seeking, and not some unknown plan of creation, or the enunciation of general propositions, and the mere putting together and separating objects more or less alike.

The diagram below shows how readily a classification scheme such as that above can be interpreted in terms of an evolutionary tree.

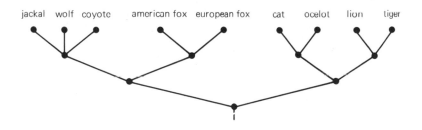

more restricted groups. The smallest division usually employed is the *species*. This term has been applied to a reproductively isolated population of organisms that possess in common one or more distinctive feature and reproduce their characteristics in fertile young. Estimates of the number of known species vary greatly, depending upon the way in which the investigator defines the term. A group of closely related species comprise a *genus*, a group of genera constitutes a *family*, related families are grouped into *orders*, related orders into *classes*, and classes into *phyla* – the primary divisions of the animal and plant kingdoms. The subject of classification is taken up more fully in Chapter Six.

It is clear that Darwin considered the phenomenon of homology, the hierarchic pattern of nature, and many other aspects of comparative anatomy as providing very powerful circumstantial evidence for his macroevolutionary claims. Such facts he wrote:[22]

> . . . proclaim so plainly that the innumerable species; genera and families, with which this world is peopled, are all descended, each within its own class or group, from common parents, and have all been modified in the course of descent, that I should without hesitation adopt this view even if it were unsupported by other facts or arguments.

Paleontology also provided Darwin with evidence that evolution had occurred. The fossil record revealed that the history of life on earth was overall one of progress from simple to more complex types of life. The first organisms to appear in the fossil record are simple invertebrates and simple plants such as seaweeds and algae; later the more complex vertebrates appear – fish first, then amphibians, followed by reptile and mammals. Moreover, even within particular groups, the more specialized types tended to occur later. Such a general succession was just what would be expected if evolution had in fact occurred.

Another piece of paleontological evidence which struck Darwin as suggestive of macroevolution was the succession of related types over periods of millions of years within geographically isolated areas of continental dimensions. Examples of this phenomenon cited in the *Origin* were the remarkable succession of marsupial species in Australia, and the development of the exotic and unique mammalian fauna in South America while that continent was isolated from North America for forty million years until the Isthmus of Panama was reformed about five million years ago:[23]

> On the theory of descent with modification, the great law of the long enduring, but not immutable, succession of the same types within the same areas, is at once explained; for the inhabitants of each quarter of the world will obviously tend to leave in that quarter, during the next succeeding period of time, closely allied though in some degree modified descendants.

The new materialistic and evolutionary view of nature presented in the *Origin* was the absolute antithesis of the previous creationist view. While, according to creationist concepts, the design of life and the adaptive complexity of every type was the result of purposeful activity on the part of God, according to Darwin it was ultimately the result of an entirely random process. The creationist position was that living nature was a static discontinuous system where the hierarchy of living beings was divided by fundamental and unbridgeable divisions into a number of types, but Darwin's evolutionary theory saw the organic world as basically a continuum of ever changing forms, where variation is radical and always tending towards creative innovation.

Darwin himself saw no limit to the extent of evolutionary change or to the power of natural selection to mould even the most complex of adaptations. At the end of the *Origin* he does not shrink from the ultimate implication that all life had evolved from a common source. Although he does not extend his theory in the *Origin* to include the origin of life, the possibility that life's emergence could also be explained in naturalistic evolutionary terms had occurred to him. In an often-quoted passage he speculates on the origin of living systems from a warm solution of organic compounds through a succession of increasingly more complex chemical aggregates:[24]

> It is often said that all the conditions for the first production of a living organism are now present which could ever have been present. But if (and oh! what a big if!) we could conceive in some warm little pond with all sorts of ammonia and phosphoric salts, light, heat, electricity present that a protein was formed ready to undergo still more complex changes at the present day such matter would be instantly devoured or absorbed which would not have been the case before living creatures were formed.

Darwin never claimed his theory could explain the origin of life, but the implication was there. Thus, not only was God banished from the creation of species but from the entire realm of biology.

Although Darwin had nearly all the key ideas of the *Origin* clear in his mind as early as 1838, he deliberated for twenty years before committing himself publicly to evolution. A number of factors were probably responsible for this delay. One may have been its controversial and anti-religious character. We know that his wife, with whom he was very close, found his views disturbing as they seemed to her to "be putting God further and further off". Darwin himself was perfectly aware of the tremendous implication of the claims of the *Origin* and particularly of the concept of evolution by natural selection which eliminated the hand of God from the design of life. His own religious beliefs had been gradually eroding as his belief in evolution had grown, and, as a sensitive person, he must have seen that the elimination of meaning and purpose from human existence, which was the inescapable conclusion of his position, was for many, including his wife, a profoundly disturbing reality to accept. Something of the personal agonies with which he may have had to come to terms is hinted at in the touching letter written by his wife. In it she appeals to him in the gentlest way possible to try to understand her religious sensibilities and difficulties with his theory.[25]

> The state of mind that I wish to preserve with respect to you is to feel that while you are acting conscientiously and sincerely wishing and trying to learn the truth, you cannot be wrong, but there are some reasons that force themselves upon me, and prevent myself from being always able to give myself this comfort. I dare say you have often thought of them before, but I will write down what has been in my head, knowing that my own dearest will indulge me.
>
> Your mind and time are full of the most interesting subjects and thoughts of the most absorbing kind, (viz. following up your own discoveries) but which make it very difficult for you to avoid casting out as interruptions other sorts of thoughts which have no relation to what you are pursuing, or to be able to give your whole attention to both sides of the question . . .
>
> It seems to me also that the line of your pursuits may have led you to view chiefly the difficulties on one side, and that you have not had time to consider and study the chain of difficulties on the other; but I believe you do not consider your opinion as formed.
>
> May not the habit in scientific pursuits of believing nothing till it is proved, influence your mind too much in other things which cannot be proved in the same way, and which if true, are likely to be above our comprehension? . . .
>
> I do not know whether this is arguing as if one side were true and the other false, which I meant to avoid, but I think not. I do not quite agree with you in what you once said that luckily there were no doubts as to how one ought to act. I think prayer is an instance to the contrary, in one case it is a positive duty and perhaps not in the other.
>
> But I dare say you meant in actions which concern others and then I agree with you almost if not quite. I do not wish for any answer to all this – it is a satisfaction to me to write it, and when I talk to you about it I cannot say exactly what I wish to say, and I know you will have

patience with your own dear wife. Don't think that it is not my affair and that it does not much signify to me. Everything that concerns you concerns me, and I should be most unhappy if I thought we did not belong to each other forever. . . .

Some idea of the conflicts in Darwin's own mind are indicated by what he wrote at the end of the letter:

> When I am dead, know that many times I have
> kissed and cried over this.
>
> C.D.

Such remarks belie the picture often painted of Darwin as a hard-headed scientific agnostic. On the contrary, if anything, he was a reluctant advocate of his views and very far from the ruthless crusader against religion and religious obscurantism as were so many of his followers, such as that arch anti-cleric, Thomas Huxley.

Darwin was not only a man of great personal sensibilities but he was also a man of great integrity, especially in scientific matters. He was by nature cautious and well aware that not only were his conclusions controversial but that the evidence was in many ways insufficient. He was acutely aware that the whole edifice he had constructed in the *Origin* was entirely theoretical.

By its very nature, evolution cannot be substantiated in the way that is usual in science by experiment and direct observation. Neither Darwin nor any subsequent biologist has ever witnessed the evolution of one new species as it actually occurs. Outside of direct observation the only means of providing decisive evidence for evolution is in the demonstration of unambiguous sequential arrangements in nature.

To show that any two species of organism are related in an evolutionary sense, to show for example that one species A, is ancestral to B, ie A→B or that both species have descended from a common ancestral source, ie A⤙B, it is necessary to satisfy one of the following conditions. Either *one*, to find a 'perfect' sequence of fully functional intermediate forms I^1, I^2, I^3 leading unambiguously from one species to another, ie A→I^1→I^2→I^3→B, or

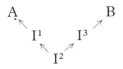

or *two*, to reconstruct hypothetically in great detail the exact sequence of events which led from A to B or from a common ancestor to A and B, including thoroughly convincing reconstructions of intermediate forms and a rigorous and detailed explanation of how and why each stage in the transformation came about.

Although incapable of providing logically compelling evidence in a formal mathematical sense, most biologists, even those opposed to evolution, have seen in the existence of clear-cut sequential patterns virtually irrefutable evidence for natural evolutionary transformations.

There is no doubt that as far as his macroevolutionary claims were concerned Darwin's central problem in the *Origin* lay in the fact that he had absolutely no direct empirical evidence in the existence of clear-cut intermediates that evolution on a major scale had ever occurred and that any of the major divisions of nature had been crossed gradually through a sequence of transitional forms. Over and over, he returns to the same problem, confessing that:[26]

> . . . the distinctness of specific forms and their not being blended together by innumerable transitional links is a very obvious difficulty

Not only was he unable to provide empirical evidence for evolution in the existence of intermediate forms, there was in many cases a real difficulty in imagining the hypothetical paths through which evolution had occurred. This was particularly true of various highly specialized organisms and organs, and Darwin concedes:[27]

> It is no doubt difficult even to conjecture by what gradations many structures have been perfected . . .
> . . . although in many cases it is most difficult even to conjecture by what transitions organs have arrived at their present state . . .[28]

No matter how suggestive the circumstantial evidence drawn from comparative anatomy, from the facts of classification and from the fossil record, to argue the case for macroevolution Darwin failed to satisfy the two basic conditions stated above. Nature remained, despite the persuasiveness of Darwin's arguments, profoundly discontinuous in both an empirical and conceptual sense.

The absence of intermediate forms essentially emptied all Darwin's macroevolutionary claims of any empirical basis. Without intermediates, not only was he unable to prove decisively that organisms had indeed evolved gradually as a result of simple random processes such as natural selection, but he had no way of distinguishing empirically between his own evolutionary model of nature and its non-evolutionary rivals, whether they were basically naturalistic, postulating sudden but natural macromutations as a basic mode of change, or frankly supernatural, invoking the intervention of God.

The challenge to his evolutionary framework was underlined by the fact that the only explanation Darwin was able to offer in the *Origin* was his appeal to the 'extreme imperfection' of the fossil record. But this was largely a circular argument because the only significant evidence he was able to provide for its 'extreme imperfection' was the very absence of the intermediates that he sought to explain.

The gaps were a particularly acute problem for Darwin as he was absolutely insistent that evolution by natural selection must be a very slow gradual process:[29]

> That natural selection generally acts with extreme slowness I fully admit . . . I do believe that natural selection will generally act very slowly, only at long intervals of time . . . Slow though the process of selection may be.[30] As natural selection acts solely by accumulating slight, successive, favourable variations, it can produce no great or sudden modifications; it can act only by short and slow steps.[31]

Such a slow gradual mechanism of evolution necessitated innumerable transitional forms and this was acknowledged freely by Darwin on many occasions in the *Origin*.

In the *Origin* Darwin was able to point to the extinct *Hipparion*,[32] an early three-toed horse, as intermediate between the existing horses and certain older five-toed forms; to the extinct Dugong *Halitherium* as intermediate between the modern Sirenia and hoofed quadrupeds because it possessed an ossified thigh bone which articulated to a well-defined acetabulum in the pelvis; to *Zeuglodon*, an early whale, as a connecting link between the Carnivora and Cetacea; and also to the *Archaeopteryx* as intermediate between modern birds and reptiles.

But none of the above examples, except in the case of *Hipparion*, are particularly convincing intermediates and unfortunately the gap between *Hipparion* and the modern horse is essentially trivial (after all, mutant horses with three toes are occasionally born today). The gap between the primitive sea cow *Halitherium* and the hoofed quadrupeds is enormous, as is the gap between the primitive whale *Zeuglodon* and the carnivores. As to *Archaeopteryx*, although it had certain reptilian characteristics, its wing possessed normal flight feathers and may have been as capable of powered flight as a modern pigeon or crow (see Chapter Eight). *Archaeopteryx* was probably the best intermediate that Darwin was able to name, yet between reptiles and *Archaeopteryx* there was still a very obvious gap.

Moreover, Darwin's theory required not just one or two intermediates of dubious status but 'innumerable' transitional forms and the fossil record provided no evidence for believing that this infinitude of connecting links had ever existed. In the *Origin* he wrote:[33]

> Geological research, though it has added numerous species to existing and extinct genera, and has made the intervals between some few groups less wide than they otherwise would have been, yet has done scarcely anything in breaking the distinction between species, by connecting them together by numerous, fine, intermediate varieties; and this not having been affected, is probably the gravest and most obvious of all the many objections which may be urged against my views.

Although the absence of intermediates was acknowledged as an enormous difficulty, Darwin never weakened in his insistence that evolution must be a gradual process. For Darwin the term evolution, which literally means 'a rolling out', always implied a very slow gradual process of cumulative change (a view which has been subscribed to by the great majority of biologists ever since). There were two main reasons why Darwin rejected the saltational solution to the challenge of the great gaps in nature. Firstly, he considered it axiomatic that all natural processes always must conform to the principle of continuity. In his book *Darwin on Man*, Howard Gruber remarks:[34]

> Natura non facit saltum – nature makes no jumps – was a guiding motto for generations of evolutionists and proto-evolutionists. But Darwin encountered it in a sharp and interesting form, posed as an alternative of terrible import: nature makes no jumps, but God does. Therefore, if we want to know whether something that interests us is of natural origin or supernatural, we must ask: did it arise gradually out of that which came before, or suddenly without any evident natural cause?

We can, of course, ask this question about anything in the natural world. We can also ask it about the very idea of God. And it was in this form that Darwin encountered the question, while a student at Cambridge. Among the pages of his student notes that survive, there are a few sheets outlining the argument of *The Evidence of Christianity Derived from its Nature and Reception* by John Bird Sumner, then Bishop of Salisbury, later to become Archbishop of Canterbury. . . . Sumner's central argument rests on a simple proposition cast in a specific logical form: nature makes no jumps, therefore if something is found in the world that appears suddenly, its origins must be supernatural. . . . Darwin made a chapter-by-chapter outline of Sumner's *Evidence*. Among his notes there is the following passage: "When one sees a religion set up, that has no existing prototype . . . it gives great probability to its divine origin."

In other words, sometime in his Cambridge years, 1827–30, Darwin took cognizance of the proposition that in order to show something is of natural origin it must be shown that it evolved gradually from its precursors, otherwise its origins are supernatural. This formulation of the choices open to rational men remained a leitmotif throughout his life.

On top of this *a priori* reason for rejecting jumps, Darwin found empirical reasons in the very great adaptive complexity of living organisms, particularly the high level of coadaptation of all their parts, "that perfection of structure and coadaptation which justly excites our admiration". Darwin, like all other biologists in the early nineteenth century, was greatly influenced and impressed by the ideas of natural theology which emphasized the ingenuity and elegance of biological adaptation. In the *Origin* he summarizes his rejection of saltationalism:[35]

He who believes that some ancient form was transformed suddenly through an internal force or tendency . . . will further be compelled to believe that many structures beautifully adapted to all the other parts of the same creature and to the surrounding conditions, have been suddenly produced; and of such complex and wonderful co-adaptations, he will not be able to assign a shadow of an explanation. . . . To admit all this is, as it seems to me, to enter into the realms of miracle, and to leave those of Science.

In the context of the great discontinuities of nature and the absence of transitional forms, Darwin's insistence that a natural evolutionary process must be infinitely gradual, requiring myriads of transitional forms, only tended to emphasize the highly theoretical nature of his claims.

Even if Darwin had been able to provide clear evidence for continuity on a grand scale, this would have still left him with the tremendous task of justifying the second great axiom of his evolutionary theory, the radical claim, that the driving force behind the whole of evolution was the purely random mechanism of natural selection.

To begin with, even by Darwin's own admission, evolution by natural selection is bound to be a relatively slow process and the question obviously arises, has there been sufficient time for all the enormous changes that must necessarily have occurred during the course of evolution? Consider the evolution of A into B through a number of mutational intermediates, ie

$$A \rightarrow I^1 - I^2 - I^3 - I^4 - I^5 - - - - - - - - I^N \rightarrow B$$
$$\longleftarrow \text{ time T } \longrightarrow$$

Each new advantageous mutation or innovation, ie $A \rightarrow I^1$, $I^1 \rightarrow I^2$, must first occur, and then spread by interbreeding to all the members of the species and the rate at which this occurs, the substitution rate, depends on a number of factors, including mutation rate, generation time and total population number★. Unless the advantageous mutation rate, the substitution rate and the total number of advantageous mutations are known, then it is simply impossible to assess whether the transition $A \rightarrow B$ could have possibly occurred by natural selection in the time available. Unfortunately the *Origin* provides no quantitative evidence of this kind to show that any one major evolutionary transformation would in fact have been possible in the manner envisaged by Darwin.

Moreover it is one thing to show that an evolutionary route is *possible* in the time available, quite another to show that it is also *probable*. Take the case of the eye, for example. Even if Darwin had been able to demonstrate the existence of a continuous sequence of increasingly complex organs of sight, leading in tiny evolutionary steps from the simplest imaginable photosensitive spot to the perfection of the vertebrate camera eye in a single phylogenetic line (in fact, no such series exists in any known lineage) and even if he had been

★Even today the rate at which substitutions can be made by natural selection is controversial. In 1859, before the nature of the gene was understood and before the advent of modern population genetics, such calculations were virtually impossible.

able to show by quantitative estimates that the immense number of mutational steps could have occurred and been substituted by natural selection in the time available, this would only have meant that evolution by natural selection was *possible*. It would not have meant that it was *probable*.

Evolution by natural selection is, as stressed above, in essence merely a special case of problem solving by trial and error. This implies that every evolutionary route followed during the course of evolution to every adaptive end must have been initially discovered and traced out as the result of a process which is in the end nothing more nor less than a gigantic random search.

While it is easy to accept that a random search might hit on mutational routes leading to relatively trivial sorts of adaptive ends, such as the best coloration for a stoat or ptarmigan or the most efficient beak forms for each of the different species of Galapagos finch. But as to whether the same blind undirected search mechanism could have discovered the mutational routes to very complex and ingenious adaptations such as the vertebrate camera eye, the feather, the organ of corti or the mammalian kidney is altogether another question. To common sense it seems incredible to attribute such ends to random search mechanisms, known by experience to be incapable, at least in finite time, of achieving even the simplest of ends. Darwin himself was often prone to self doubt over the sheer enormity of his own claims:[36]

> Although the belief that an organ so perfect as the eye could have been formed by natural selection, is enough to stagger any one . . .
> I have felt the difficulty far too keenly to be surprised at others hesitating to extend the principle of natural selection to so startling a length.[37]

The only way Darwin could have countered these doubts would have been by the provision of rigorous quantitative evidence in the form of probability estimates to show that the routes to such seemingly remarkable ends could have been found by chance in the time available. To have estimated the probability that a purely random search would have discovered the route (or routes) to the eye, for example, he would have needed to have mapped out all possible routes that evolution might conceivably have taken from the original light sensitive spot over the past three thousand million years and then to have determined the fraction of routes which lead to "camera type" eyes and the fraction which lead to all other less sophisticated organs of sight. Only then would he have been able to counter his critics with quantitative evidence that such seemingly improbable ends could have been hit on by chance.

The provision of such estimates is of course the correct procedure in any area to justify a claim that phenomena are the result of chance. Yet nowhere in the *Origin* is any attempt made to provide quantitative support for the grand claim of the all-sufficiency of chance.

It is true that Darwin appealed on many instances in the *Origin* to the enormous periods of time available to the evolutionary search but, as is the case with any other random search procedure, time in itself tells us nothing of the probability of achieving any sort of goal unless the complexity of the search can be quantified.

Altogether the problems that Darwin faced in defending his general theory are underlined by the fact that he was forced to devote a large portion of the book to attempting to explain away much evidence which was on the face of it and by his own admission hostile to the whole evolutionary picture. Even the chapter titles in the *Origin* (titles such as "Difficulties of the Theory", "The Imperfections of the Fossil Record" and "Miscellaneous Objections to the Theory of Natural Selection") illustrate how seriously he took the problems he faced.

It was not only his general theory that was almost entirely lacking in any direct empirical support, but his special theory was also largely dependent on circumstantial evidence. A striking witness to this is the fact that nowhere was Darwin able to point to one *bona fide* case of natural selection having actually generated evolutionary change in nature, let alone having been responsible for the creation of a new species. Even in the case of trivial adaptations Darwin was forced to use conditional language:[38]

> When we see leaf-eating insects green, and bark feeders mottled-grey; the alpine ptarmigan white in winter, the red-grouse the colour of heather, *we must believe* that these tints are of service to these birds and insects in preserving them from danger! Grouse, if not destroyed at some period of their lives, would increase in countless numbers; they are known to suffer largely from birds of prey; and hawks are guided by eye-sight to their prey – so much so, that on parts of the Continent persons are warned not to keep white pigeons, as being the most liable to destruction. Hence natural selection *might* be effective in giving the

proper colour to each kind of grouse, and in keeping that colour, when once acquired, true and constant.

[*emphasis added*]

Moreover, Darwin faced another problem with natural selection that has been largely forgotten. In the nineteenth century the theory of inheritance which was widely accepted and in which Darwin himself believed, known as blending, was very difficult, if not impossible, to reconcile with the idea of natural selection.

The term 'blending inheritance' refers to the view (now known to be quite wrong) that inheritance is basically a fusion of both paternal and maternal elements in the offspring in an inseparable mixture which results in external features which appear to be intermediate between the two. This view was merely a description based on a purely superficial observation of what appears to occur in most crosses.

According to Peter Vorzimmer, who has made a detailed study of the problem posed to Darwin by this archaic and, as it proved erroneous, theory of inheritance:[39]

> The concept of blending inheritance as a natural process, together with all its implications, has also been called "the paint-pot theory of heredity." The analogy is that of the normal population as a bucket of white paint with the variant forms a few drops of black. The result of mixing the two paints is analagous to the effect of free inter-crossing in nature. Just as it is impossible to separate two once-distinct fluids after mixing, so it is also impossible because of the heritage of blending, for small changes to be accumulated in the process of natural selection.

Darwin was never able to fully reconcile this swamping effect of blending inheritance with evolution by natural selection and the contradiction was seized upon by some of his critics. Fleming Jenkins wrote in the *North British Review* in 1867:[40]

> It is impossible that any sort of accidental variation in a single individual, however, favourable to life, should be preserved and transmitted by Natural Selection . . . (because) the advantage, whatever it may be, is utterly outbalanced by numerical inferiority . . . (such) variation would be swamped (just as) a highly favoured white cannot blanch a nation of negroes.

Darwin acknowledges the difficulty in the fifth edition of the *Origin*:[41]

> I saw . . . that the preservation in a state of nature of any occasional deviation of structure, such as a monstrosity, would be a rare event; and that, if preserved, it would generally be lost by subsequent inter-crossing with ordinary individuals. Nevertheless, until reading an able and valuable article in the *North British Review* (1867), I did not appreciate how rarely single variations whether slight or strongly-marked, could be perpetuated.

To overcome the problem of swamping, an inevitable outcome of blending inheritance, Darwin had to concede, according to Vorzimmer, that:[42]

> . . . his theory required a significant number of the necessary variations – if not simultaneously, at least within a few generations.

To account for the required number of simultaneous beneficial variations he was forced into an almost vitalistic Lamarckian position, having to toy with the idea of some sort of directional bias in the occurrence of variation and mutation.

Neither Darwin nor any other nineteenth-century biologist had any idea of the true nature of the gene and of the mechanism of inheritance. It was only in the first decade of the twentieth century, with the founding of classical Mendelian genetics and the so-called 'bean-bag' theory of heredity (as opposed to the paint pot theory), that the fundamental units of heredity, the genes, were first shown to be quite discrete elements (the beans in the bean bag) which act separately and are inherited essentially unchanged. With the advent of Mendelian genetics, biology at last possessed a model of heredity which could explain why the influence of a single new advantageous genetic trait would not be swamped by blending but, on the contrary, could, just as Darwin believed, come over several generations to influence an entire population due only to the preferential survival of all members of the population which possessed it.

There was also the disturbing point, which Darwin was well aware of and had tried rather unconvincingly to dismiss at the end of Chapter Two of the *Origin*, that while breeding experiments and the domestication of animals had revealed that many species were capable of a considerable degree of change, they also revealed distinct limits in nearly every case beyond which no further change could ever be produced. Here then was a very well established fact, known for centuries, which seemed to run counter to his whole case, threatening not only his special theory – that one species could evolve into another – but also the plausibility of the extrapolation from micro to macroevolution, which, as we have seen, was largely based on an appeal to the remarkable degree of change achieved by artificial selection in a relatively short time. If this change was always strictly limited then the validity of the extrapolation was obviously seriously threatened.

The highly speculative nature of his evolutionary model was quite apparent to Darwin himself. Although convinced of the reality of evolution, nowhere, either in the *Origin* or in any of his other writings including his autobiography and letters, is he ever dogmatic or fanatical in his advocacy. Darwin always reveals himself to be a man of great common sense convinced, but still aware of the hypothetical nature, of his theory. Considering Darwin's sensitivity, especially with regard to his wife's religious feelings, his dislike of controversy, and his scientific caution in the face of the many deficiencies in his arguments it is not surprising that he delayed publishing for so long.

Moreover Darwin was faced, unlike modern advocates of evolution, with the task of convincing an essentially sceptical audience, and unfortunately much of his evidence was neutral and capable of alternative interpretation. The evidence from classification, for example, was suggestive of some kind of theory of descent but it did not prove evolution had occurred in the way Darwin believed (see Chapter 6) and was equally capable of creationist interpretation. Its ambiguity was implied by the fact that, as Simpson points out:[43]

> . . . from their classifications alone it is practically impossible to tell whether zoologists of the middle decades of the nineteenth century were evolutionists or not.

The evidence of homology and paleontology were also only suggestive and open to alternative explanations.

Finally, however, in 1858, twenty years after his views on evolution and natural selection had first crystallized, Darwin ceased deliberating and, prompted by the knowledge that other biologists were working on the same idea, got down to composing his great work. In 1859 the first edition of *The Origin of Species* was ready for publication. On the whole, despite the many problems and its ultimately inconclusive nature, the *Origin* was a masterful work, persuasively written. It proved a sensational best seller from the start, with all the copies of the first edition sold out on the first day. It sent a shock wave through Victorian England and was the subject of all manner of histrionic attacks.

The *Origin* was revolutionary and shocking to Victorians because nineteenth-century England was steeped in biblical fundamentalism and creationist biology. The thesis Darwin had developed implied an end to the traditional and deeply held teleological and anthropocentric view of nature. Instead of being the pinnacle and end of creation, humanity was to be viewed ultimately as a cosmic accident, a produce of a random process no more significant than any one of the myriads of other species on earth.

As far as Christianity was concerned, the advent of the theory of evolution and the elimination of traditional teleological thinking was catastrophic. The suggestion that life and man are the result of chance is incompatible with the biblical assertion of their being the direct result of intelligent creative activity. Despite the attempt by liberal theology to disguise the point, the fact is that no biblically derived religion can really be compromised with the fundamental assertion of Darwinian theory. Chance and design are antithetical concepts, and the decline in religious belief can probably be attributed more to the propagation and advocacy by the intellectual and scientific community of the Darwinian version of evolution than to any other single factor.

Today ensconced in our comfortable agnosticism, after a century of exposure to the idea of evolution and quite inured to the idea of a universe without purpose, we tend to forget just what a shock wave the advent of evolution sent through the Christian society of Victorian England. Trevelyan captured something of the mood in his *Social History of England*:[44]

> More generally speaking, the whole idea of evolution and of 'man descended from a monkey' was totally incompatible with existing religious ideas of creation and of man's central place in the universe.
>
> Naturally the religious world took up arms to defend positions of dateless antiquity and prestige. Naturally the younger generation of scientific men rushed to defend their revered chief, and to establish their claim to come to any conclusion to which their researches led,

regardless of the cosmogony and chronology of Genesis, and regardless of the ancient traditions of the Church. The strife raged throughout the sixties, seventies, and eighties. It came to involve the whole belief in the miraculous, extending into the borders of the New Testament itself. The 'intellectuals' became more and more anti-clerical, anti-religious, and materialistic under the stress of the conflict.

During this period of change and strife, causing much personal and family unhappiness and many searchings of heart, the world of educated men and women was rent by a real controversy, which even the English love of compromise could not deny to exist.

It was because Darwinian theory broke man's link with God and set him adrift in a cosmos without purpose or end that its impact was so fundamental. No other intellectual revolution in modern times (with the possible exception of the Copernican) so profoundly affected the way men viewed themselves and their place in the universe.

NOTES

1. Haeckel, E. (1866) *General Morphologie der Organismen II*, George Reimer, Berlin, Plate 1.
2. Gomperz, T. (1901) *The Greek Thinkers*, 4 vols, John Murray, London, vol 1, p 244.
3. Hume D. (1779) *Dialogues Concerning Natural Religion*, Fontana Library ed (1963), Collins, London, pp 155–56.
4. Darwin, F., ed (1888) *The Life and Letters of Charles Darwin*, 3 vols, John Murray, London, vol 1, p 83.
5. Monod, J. (1972) *Chance and Necessity*, Collins, London, p 110.
6. Simpson, G. G. (1962) see Foreword, p 5, 6th ed *Origin of the Species*, Collier Books, New York.
7. Huxley, J., Dobzhansky, T., Niebuhr, R., Reiser, O. L., Nikhilananda, S. (1958) *The Book that Shook the World*, University of Pittsburgh Press, Pittsburgh.
8. Simpson, G. G., op cit, pp 7–8.
9. Darwin, C. (1872) *The Origin of Species*, 6th ed, 1962, Collier Books, New York, p 41.
10. ibid, p 25.
11. ibid, p 405.
12. ibid, p 410.
13. Mayr, E. (1970) *Populations, Species and Evolution*, Harvard University Press, Cambridge, Mass, p 10.
14. Darwin, C., op cit, p 478
15. ibid, p 112.
16. ibid, p 312.
17. ibid, p 114–15.

18. Darwin, C., op cit, p434.
19. Hardy, A.C. (1965) *The Living Stream*, Collins, London, Figure 15, p51.
20. ibid, p450.
21. ibid, p421.
22. ibid, p457.
23. ibid, p359.
24. Darwin, F., op cit, vol 3, p18.
25. Barlow, N. (1958) *Autobiolography of Charles Darwin*, Collins, London, pp235–37.
26. Darwin, C., op cit, p307.
27. ibid, p459.
28. ibid, p192.
29. ibid, p114.
30. ibid, p114.
31. ibid, p468.
32. ibid, p349.
33. ibid, p462.
34. Gruber, H. (1981) *Darwin on Man: A Psychological Study of Scientific Creativity* 2nd ed, University of Chicago Press, Chicago, pp125–26.
35. Darwin, C., op cit, p242.
36. ibid, p192.
37. ibid, p181.
38. ibid, pp94–95.
39. Vorzimmer, P.J. (1972) *Charles Darwin: The Years of Controversy*, London University Press, London, p100.
40. ibid, p122.
41. ibid, p125.
42. ibid, p125.
43. Simpson, G. G. (1945) "The Principles of Classification and a Classification of Mammals", *Bull. Am. Mus. Nat. Hist.*, vol 85, pp1–350, see p4.
44. Trevelyan, G.M. (1944) *Social History of England*, Longmans, Green and Co, London, pp565–566.

CHAPTER 3
From Darwin to Dogma

> I cannot remember a single first-formed
> hypothesis which had not after a time to
> be given up or greatly modified.

The popular conception of a triumphant Darwin increasingly confident after 1859 in his views of evolution is a travesty. On the contrary, by the time the last edition of the *Origin* was published in 1872, he had become plagued with self-doubt and frustrated by his inability to meet the many objections which had been levelled at his theory. According to Loren Eiseley:[1]

> A close examination of the last edition of the *Origin* reveals that in attempting on scattered pages to meet the objections being launched against his theory the much-laboured-upon volume had become contradictory. . . . The last repairs to the *Origin* reveal . . . how very shaky Darwin's theoretical structure had become. His gracious ability to compromise had produced some striking inconsistencies. His book was already a classic, however, and these deviations for the most part passed unnoticed even by his enemies.

There can be no question that Darwin had nothing like sufficient evidence to establish his theory of evolution. Neither speciation nor even the most trivial type of evolution had ever actually been observed directly in nature. He provided no direct evidence that natural selection had ever caused any biological change in nature and the concept was in itself flawed because it was impossible to reconcile with the theory of heredity in vogue at that time. The idea of evolution on a grand scale was entirely speculative and Darwin was quite unable to demonstrate the "infinitude of connecting links", the existence of which he repeatedly admitted was crucial to his theory. The objections were many and challenging and Darwin, because of his intellectual honesty, never tried to pretend otherwise. The difficulties always weighed heavily upon him.

Yet despite the weakness of the evidence, Darwin's theory was elevated from what was in reality a highly speculative hypothesis into an unchallenged dogma in a space of little more than twenty years after the publication of the *Origin*. To understand how this came about we have to look beyond the facts of biology. As is so often the case and as the history of science so amply testifies, the acceptance of new ideas is often dependent on the influence of non-scientific factors of a social, psychological and philosophical nature and the Darwinian revolution was no exception.

To begin with, the concepts of continuity and gradualism which were basic to the whole Darwinian model of evolution were in keeping with a general tendency towards political and social conservatism which was prevalent in nineteenth-century Victorian society and deeply ingrained in modern western societies. Stephen Jay Gould and Niles Eldredge comment:[2]

> The general preference that so many of us hold for gradualism is a metaphysical stance embedded in the modern history of Western cultures: it is not a high-order empirical observation, induced from the objective study of nature. The famous statement attributed to Linnaeus – nature non facit saltum (nature does not make leaps) – may reflect some biological knowledge, but it also represents the translation into biology of the order, harmony and continuity that European rulers hoped to maintain in a society already assaulted by calls for fundamental social change.

Another social factor which probably eased the way for Darwin was the Victorian belief in the inevitability of progress. This optimistic view of the unlimited possibilities for human progress and the belief in the perfectability of man may seem naive today but such a social evolutionary philosophy could hardly have hindered the spread and acceptance of the idea of biological evolution. Further, no society could have been more receptive to the concept of natural selection than Victorian England. Herbert Spencer was not the only intellectual at the time to draw an analogy between the competitive spirit of the free market economy as the driving force behind social and economic progress and Darwin's concept of natural selection as the driving force behind evolution.

But undoubtedly the most significant factor that contributed to the success of Darwinian theory after 1859 was the fact that it was the first genuine attempt to bring the study of life on Earth fully into the conceptual sphere of science. Sevententh and eighteenth-century explanations for physical phenomena, in terms of final causes or supernatural intervention, had been dropped from physics and re-placed by the new scientific method espoused by Bacon and Descartes and epitomized by the grand Newtonian synthesis. The aim of the new physics was to give first and natural causes for all physical phenomena. In this it had been dramatically successful. God came to be viewed increasingly as a distant and remote first cause, the architect of a clockwork universe which had continued from its creation to operate automatically without any need for further divine inter-vention. Explanations in terms of natural causes were also having increasing success in chemistry and physiology. Wohler's artificial synthesis of urea revealed that the formation of organic compounds, previously held to be the result of some mysterious vitalistic force, was also explicable in terms of normal chemical processes.

It seemed increasingly likely to most educated men of the time that all past phenomena would prove explicable in terms of presently operating processes and that the universe had gradually developed from a few elementary particles into its present state through the operation of the basic laws of physics and chemistry. Thus, the universe came to be viewed in uniformitarian terms as a closed system and all the phenomena within it as essentially natural. Since its origin it had suffered no unnatural peturbation or interference.

By the mid-nineteenth century it was apparent to everyone – even the most rigid religious dogmatists – that everyday physical phenom-ena were reducible and readily explicable in terms of natural causes. The miraculous interventionist world view of the middle ages seemed for less credible. What Darwin was doing therefore in the *Origin* was extending the scientific method to the biological sciences by giving a natural explanation of the design of living things.

In the decades before 1859 uniformitarian concepts had been applied to geological phonomena by Lyell. Lyell had shown that the great geological upheavals of the past, and the enormous changes during the long geological history of the Earth, were all easily explic-able in terms of well-understood processes which were occurring today. After Lyell it was felt no longer necessary to propose, as had Cuvier and the Catastrophists, a series of supernatural interventions in the history of the planet to explain the sudden extinction of whole nas and the uplift of sediments deposited beneath the seas to the top of mountains. The success of Lyell's application of uniformitarian thinking is known to have impressed Darwin.

Biological evolution was the natural and inevitable consequence of extending uniformitarian thinking into biological sciences. This was admitted by the advocates of evolution in the years following 1859. As Huxley confessed in 1887:[3]

> I have recently read afresh the first edition of the "Principles of Geology;" and when I consider that for nearly thirty years this remarkable book had been in everybody's hands, and that it brings home to every reader of ordinary intelligence a great principle and a great fact – the principle that the past must be explained by the present unless good cause can be shown to the contrary, and the fact that, so far as our knowledge of the past history of life on our globe goes, no such cause can be shown – I can not but believe that Lyell was, for others, as for myself, the chief agent in smoothing the road for Darwin. For consistent uniformitarianism postulates evolution as much in the organic as the inorganic world. The origin of a new species by other than ordinary agencies would be a vastly greater "catastrope" than any of those which Lyell successfully eliminated from sober geological speculation.

An appeal to uniformitarian ideas was a favourite line of argument of evolutionists after 1859. Professor Tyndall, in his Belfast Address in 1874, pointed out that:[4]

> ... the basis of the doctrine of evolution consists, not in an experimental demonstration – for the subject is hardly accessible to this mode of proof – but in its general harmony with scientific thought. . . . We claim and we shall wrest from theology, the entire domain of cosmological theory. All schemes and systems which thus infringe upon the domain of science must, in so far as they do this, submit to its control. . . . Acting otherwise has always proved disastrous in the past, and it is simply fatuous today.

Likewise in 1882 Romanes put in the forefront of the arguments for evolutionism:[5]

> No one ever thinks of resorting to supernaturalism, except in the comparatively few cases where science has not yet been able to explore the most obscure regions of causation. . . . We are now in possession of so many of these historical analogies, that all minds with any instincts

of science in their composition have grown to distrust, on merely antecedent grounds, any explanation which embodies a miraculous element. . . . Now, it must be obvious to any mind which has adopted this attitude of thought, that the scientific theory of natural descent is recommended by an overwhelming weight of antecedent presumption.

In effect, there was no scientific alternative. If nature was fundamentally discontinuous, as typology maintained, it was far harder, if not impossible, to envisage what sort of natural process could have generated all the diversity of life on earth. As we saw in the previous chapter, Darwin held the axiom *natura non facit saltum* to be an unbreakable rule of nature. For him, as for most biologists ever since, nature simply cannot be discontinuous while at the same time reducible to natural explanations. Furthermore, none of the typologists had even attempted to provide a scientific explanation of how the discontinous pattern they perceived could have come about.

Moreover, typology had a frankly metaphysical basis; it seemed increasingly dated and was associated, as mentioned in Chapter One, with the traditional biblical framework, with teleology, special creationism and indeed with the miraculous interventionist medieval world view. Thus typology was damned by association as well as by the fact that fundamental discontinuities seemed irreconcilable with naturalistic explanations.

To Darwin and most of his contemporaries, particularly in the English speaking world, the only alternative seemed to be a very narrow type of special creationism which was not only unscientific but had also been discredited by the fact that the species barrier seemed to have been breached by perfectly natural processes.

When the appeal of the scientific paradigm and the natural desire of the scientific community to extend the range of scientific explanation are taken in conjunction with all the various intellectual trends and fashions of the later Victorian era, it is in retrospect perfectly easy to understand how Darwin's theory proved irresistible even though, as Darwin himself admitted, the actual empirical evidence was insufficient, and there was absolutely no evidence that any of the major divisions of nature had been crossed in a gradual manner. If nature was to be explained by natural processes, she had to be continuous.

In the wake of the great debate which followed the publication of *The Origin of Species*, the majority of biologists (apart from some notable exceptions such as Owen and Agassiz) came to view life as the product of a natural evolutionary process. But this does not mean that the Darwinian model was established as a fact. What the decades following 1859 witnessed was a dramatic overthrow of one particular interpretation of nature and its replacement by an entirely antithetical theory. Biologists adopted a quite new framework through which to visualize nature. Before 1859 it was fashionable and intellectually respectable to view the organic world as a discontinuous system – the result of successive creative interventions in the history of the world. After 1859 it became intellectually respectable to view life as the natural product of purely natural processes operating over long periods of time. Changing one's interpretation of the world is not, however, the same as establishing a new fact. The facts were the same in 1850 as they were in 1870, only the perception of them had changed. What had happened was something akin to the change in perception that occurs when viewing a reverse figure diagram or Escher engraving. First we see one pattern, then later a quite different pattern is perceived, but the picture remains the same.

Philosophers and historians of science will probably be debating the nature of the Darwinian revolution for years to come, but whatever their final verdict on this event, the facts themselves were not sufficient to compel belief in the continuity of living nature or to establish beyond reasonable doubt that the whole drama of life on earth was generated by the sorts of simple random processes responsible for microevolution on the Galapagos Islands.

As the years passed after the Darwinian revolution, and as evolution became more and more consolidated into dogma, the gestalt of continuity imposed itself on every facet of biology. The discontinuities of nature could no longer be perceived. Consequently, debate slackened and there was less need to justify the idea of evolution by reference to the facts.

Increasingly, its highly theoretical and metaphysical nature was forgotten, and gradually Darwinian concepts came to permeate every aspect of biological thought so that today all biological phenomena are interpreted in Darwinian terms and every professional biologist is subject throughout his working life to continued affirmation of the truth of Darwinian theory.

The fact that every journal, academic debate and popular discussion assumes the truth of Darwinian theory tends to reinforce its credibility enormously. This is bound to be so because, as sociologists of knowledge are at pains to point out, it is by conversation in the broadest sense of the word that our views and conceptions of reality are maintained and therefore the plausibility of any theory or world view is largely dependent upon the social support it recieves rather than its empirical content or rational consistency. Thus the all pervasive affirmation of the validity of Darwinian theory has had the inevitable effect of raising its status into an impregnable axiom which could not even conceivably be wrong.

It is not surprising that, in the context of such an overwhelming social consensus, many biologists are confused as to the true status of the Darwinian paradigm and are unaware of its metaphysical basis. As the following quote from Julian Huxley at a conference in 1959 makes clear:[6]

> The first point to make about Darwin's *theory* is that it is no longer a *theory* but a *fact* . . . Darwinianism has come of age so to speak. We are no longer having to bother about establishing the *fact* of evolution . . .
>
> [*emphasis added*]

Richard Dawkins, author of *The Selfish Gene*, is even more emphatic, for him:[7]

> The theory is about as much in doubt as the earth goes round the sun.

Now of course such claims are simply nonsense. For Darwin's model of evolution is still very much a theory and still very much in doubt when it comes to macroevolutionary phenomena. Furthermore being basically a theory of historical reconstruction, it is impossible to verify by experiment or direct observation as is normal in science. Recently the philosophical status of evolutionary claims has been the subject of considerable debate. Philosophers such as Sir Karl Popper have raised doubts as to whether evolutionary claims, by their very nature incapable of falsification, can properly be classed as truly scientific hypotheses. Moreover, the theory of evolution deals with a series of unique events, the origin of life, the origin of intelligence and so on. Unique events are unrepeatable and cannot be subjected to any sort of experimental investigation. Such events, whether they be the origin of the universe or the origin of life, may be the subject of much fascinating and controversial speculation, but their causation can, strictly speaking, never be subject to scientific validation.

Furthermore, not only is the theory incapable of proof by normal scientific means, the evidence is, as we shall see in the next few chapters, far from compelling.

Although it is nonsense to claim that Darwin's theory is a fact, ironically both Huxley and Dawkins are right in the sense that, once a community has elevated a theory into a self-evident truth, its defence becomes irrelevant and there is no longer any point in having to establish its validity by reference to empirical facts

The transformation of Darwinian theory into dogma is evidenced also by the hostility that is directed towards the dissidents from orthodoxy such as Klammerer in the 1920's[8] and recently the Australian geneticist Steel[9] for raising the possibility of Lamarckianism, and towards authorities such as the geneticist Goldschmidt[10] and the paleontologist Schindewolf[11] for rejecting natural selection as the major agency in macroevolution. Such hostility is readily understandable in terms of the sociology of knowledge because, as the biological community considers Darwinian theory to be established beyond doubt "like the earth goes round the sun", then dissent becomes by definition irrational and hence especially irritating if the dissenters claim to be presenting a rational critique. It is ironic to reflect that while Darwin once considered it heretical to question the immutability of species, nowadays it is heretical to question the idea of evolution.

Once a theory has become petrified into a metaphysical dogma it always holds enormous explanatory power for the community of belief. As Paul Feyerabend explains:[12]

> The conceptual apparatus of the theory and the emotions connected with its application having penetrated all means of communication, all actions, and indeed the whole life of the community, such methods as transcendental deduction, analysis of usage, phenomenological analysis which are means for further solidifying the myth will be extremely successful. . . . Observational results, too, will speak in favour of the theory as they are formulated in its terms. It will seem that at last the truth has been arrived at . . . .

This semblance of truth is of course a mirage, as Feyerabend continues:[13]

> . . . the stability achieved, the semblance of absolute truth is nothing but the result of an absolute conformism. For how can we possibly

test, or improve upon, the truth of a theory if it is built in such a manner that any conceivable event can be described, and explained, in terms of its principles? The only way of investigating such all-embracing principles is to compare them with a different set of equally all-embracing principles – but this way has been excluded from the very beginning. The myth is therefore of no objective relevance, it continues to exist solely as the result of the effort of the community of believers and of their leaders, be these now priests or Nobel prize winners. Its "success" is entirely man made.

The raising of the status of Darwinian theory to a self-evident axiom has had the consequence that the very real problems and objections with which Darwin so painfully laboured in the *Origin* have become entirely invisible. Crucial problems such as the absence of connecting links or the difficulty of envisaging intermediate forms are virtually never discussed and the creation of even the most complex of adaptations is put down to natural selection without a ripple of doubt.

The overriding supremacy of the myth has created a widespread illusion that the theory of evolution was all but proved one hundred years ago and that all subsequent biological research – paleontological, zoological and in the newer branches of genetics and molecular biology – has provided ever-increasing evidence for Darwinian ideas. Nothing could be further from the truth. The fact is that the evidence was so patchy one hundred years ago that even Darwin himself had increasing doubts as to the validity of his views, and the only aspect of his theory which has received any support over the past century is where it applies to microevolutionary phenomena. His general theory, that all life on earth had originated and evolved by a gradual successive accumulation of fortuitous mutations, is still, as it was in Darwin's time, a highly speculative hypothesis entirely without direct factual support and very far from that self-evident axiom some of its more aggressive advocates would have us believe.

NOTES

1. Eisley, L. (1959) *Darwin's Century*, Gollancz, London, p 242.
2. Gould, S.J. and Eldrige, N. (1977) "Punctuated Equilibria: The Tempo and Mode of Evolution Reconsidered", *Paleobiology*, 3: 115–51, see p 145.
3. Huxley, T.H. (1887) "Science and Pseudoscience" in A.O. Lovejoy (1959) "The Argument for Organic Evolution before the *Origin of Species*, 1830–1858"

in *Forerunners of Darwin: 1745–1859*, eds B. Glass, O. Temkin and W. L. Strauss, pp356–414, see p374

4. Tyndall, J. (1874) *Presidential Address to British Association*, George Robertson, Melbourne, pp43–44.

5. Romanes, G. F. (1882) cited in Lovejoy, op cit, p376.

6. Huxley, J. (1960) "The Emergence of Darwinism" in *Evolution of Life*, ed Sol Tax, University of Chicago Press, Chicago, pp1–21, see p1.

7. Dawkins, R. (1976) *The Selfish Gene*, Oxford University Press, p1.

8. Koestler, A. (1971) *The Midwife Toad*, Hutchinson Publishing Co, London.

9. Editorial (1981) "Too Soon for the Rehabilitation of Lamark", *Nature*, 289: 631–32.

10. Goldschmidt, R. (1940) *The Material Basis of Evolution*, Yale University Press, New Haven.

11. Schindewolf, O. H., cited in B. Rensch (1959) *Evolution above the Species*, Columbia University Press, New York, p58.

12. Feyerabend, P. (1965) "Problems of Empiricism" in *Beyond the Edge of Certainty*, ed R. G. Colodny, pp145–260, see p176.

13. ibid, p179.

CHAPTER 4
A Partial Truth

Grouse, if not destroyed at some period of their lives, would increase in countless numbers; they are known to suffer largely from birds of prey; and hawks are guided by eyesight to their prey – so much so, that on parts of the Continent persons are warned not to keep white pigeons, as being the most liable to destruction. Hence natural selection might be effective in giving the proper colour to each kind of grouse, and in keeping that colour, when once acquired, true and constant.

It is a striking testimony to the fundamentally theoretical character of Darwin's arguments in the *Origin* that none of his claims received any direct experimental support until nearly a century had elapsed. Biology had to wait until the early 1950s and the work of the Oxford zoologist Bernard Kettlewell for clear evidence that natural selection actually operated in nature. Before 1950 it was still possible to argue that selection only occurred in artificial situations such as in the domestication of animals and the production of particular breeds by man. The following statement by the ecologist Cyril Diver made at a special symposium organized by the Royal Society in 1936 entitled "The Present State of the Theory of Natural Selection" illustrates the total lack of empirical evidence for Darwin's ideas at that time:[1]

> The degree to which Natural Selection is responsible for the production of adaptations and the origin of species is still a matter which must be discussed largely on indirect evidence. Recent genetical and mathematical work has greatly increased the reasonableness of the theory, but it has mainly been directed to demonstrating that Natural Selection could operate and not that it does operate. ...

It was a common species of moth – the peppered moth – which provided Kettlewell with the first clear evidence that natural selection did in fact operate in nature. It was well known that a century ago the British peppered moth was light-coloured throughout its range. This light coloration matched such backgrounds as light-coloured trees and lichen covered rocks on which the moths passed the day at rest. Yet today in most industrial regions where the trees and rocks are darkened by industrial pollution the same species of moth is predominantly dark in colour. The dark form is far better camouflaged in its industrial environment than the light. Kettlewell set out to answer the question of why the dark form had become so common in industrial regions of England over the past century that it almost completely eclipsed the light form in certain areas.

By a simple series of experiments Kettlewell was able to show that in polluted areas, where the background was dark, the dark forms were far less visible to their major predator – birds – and survived far better than their light-coloured cousins, while in rural areas the situation was reversed with the light forms surviving far better than the dark. In Kettlewell's own words:[2]

> We decided to test the rate of survival of the two forms in the contrasting types of woodland. We did this by releasing known numbers of moths of both forms. Each moth was marked on its underside with a spot of quick-drying cellulose paint; a different color was used for each day. Thus when we subsequently trapped large numbers of moths we could identify those we had released and establish the length of time they had been exposed to predators in nature.
>
> In an unpolluted forest we released 984 moths: 488 dark and 496 light. We recaptured 34 dark and 62 light, indicating that in these woods the light form had a clear advantage over the dark. We then repeated the experiment in the polluted Birmingham woods, releasing 630 moths: 493 dark and 137 light. The result of the first experiment was completely reversed; we recaptured proportionately twice as many of the dark form as of the light.

Kettlewell's work attracted widespread interest and one of his colleagues at Oxford, Professor Niko Tinbergen, went on to actually film the birds capturing and eating the moths, selectively choosing light moths on the darkened tree trunks and dark ones in the rural areas where the tree trunks were light. Where Darwin had argued that natural selection *might* work, now at last it could be claimed that it did. And it is not only the coloration of moths that has now been shown to be under the control of natural selection; since Kettlewell's pioneering work several other cases of natural selection have been directly observed.

Although impressive, the sort of evolution observed by Kettlewell is relatively trivial and falls far short of the evolution of a new species if we accept the usual definition of a species as a reproductively isolated population of organisms. Even today, the origin of a new species from a pre-existing species has never been directly observed. This is not surprising since, as Mayr points out,[3] "speciation is a slow historical process, it can never be directly observed by an individual observer."

As noted in Chapter Two there are only two sorts of evidence for evolution which do not depend on actual observation of the process: finding a sequence of intergrading forms leading unambiguously from one form to another; or reconstructing them hypothetically by providing an entirely plausible genealogy including all the inter-mediate forms and a thoroughly convincing explanation of how each stage of the transformation came about. Although Darwin was never able to provide this sort of evidence to support his claim that speciation had actually occurred in nature, over the past few decades both approaches have been applied extensively with the result that the fact of speciation in nature can hardly be doubted.

One of the clearest examples of the first kind of evidence is presented by the phenomenon of circular overlaps. This is where there is a chain of intergrading subspecies forming a loop or overlapping circle whose terminal links, although inhabiting the same geographical region, do not interbreed even though they are connected by a complete chain of interbreeding populations.

The classic case of this is the two species of European gull – the herring gull (*Larus argentatus*) and the lesser black backed gull (*Larus fuscus*).[4] In Europe the two species are distinct. They do not interbreed and are quite different in terms of appearance and behav-iour. However, if one goes east across Russia and Siberia the herring gull does not occur and the lesser black backed gull becomes increas-ingly unlike the European type and comes gradually, by the time one reaches the Bering Straits, to resemble the herring gull. Similarly, if one travels west across the Atlantic the lesser black back does not occur but the herring gull, which is found right across the northern regions of North America, increasingly comes to resemble the lesser black backed gull the farther west one goes. In Eastern Siberia there is a form of gull which is almost exactly intermediate between the herring gull and the lesser black backed gull. All the different races interbreed with adjacent races except at the two ends of the ring where the two forms are two distinct non-interbreeding species. One can trace, step by step, the formation of the two species by following the intergrading subspecies right round the northern hemisphere. A more dramatic demonstration of the reality of speciation in nature can hardly be imagined!

Four thousand miles northwest of the Galapagos Archipelago are the islands of Hawaii, which present a very similar overall picture to that of the Galapagos Islands except that their fauna is richer and contains many more unique species which have diversified to a much greater extent. Insect evolution on Hawaii has been particularly spectacular. There are about 4300 species of insects unique to the Hawaiian Archipelago and these appear to have descended from about 250 original colonizations.[5] In some cases single colonizations have resulted in hundreds of descendant species. For example, from only one or two original colonizations by the fruit fly *Drosophila*, something like six or seven hundred unique Hawaiian species have evolved. Similarly, single colonizations by snails, moths, beetles and wasps have resulted in the development of large numbers of derived species, all of which are unique to Hawaii.

Study of the *Drosophila* species of Hawaii has revealed one of the most dramatic cases where "perfect" sequential arrangements have provided compelling evidence that new species do arise from pre-existing species in nature.

Fruitflies and other flies possess grossly enlarged chromosomes in certain organs such as their salivary glands, and by special techniques the arrangements of the genes along these giant chromosomes can be visualized. By studying the order of genes along the chromosomes in the various *Drosophila* species on the different islands biologists have found a number of perfect evolutionary sequences and have been able to work out the entire evolutionary history of most of the Hawaiian species.[6] The sort of evolutionary series found in the gene arrangement, in the chromosomes of the fruitflies is illustrated in the three strings of letters shown below, where each string represents the sequence of genes in a large chromosome in a particular species:

	(1)	(2)	(3)
	C	C	C
	B	B	B
	A	A	D
	B	B	B
	B	B	B
	D	D	A
	F	E	E
	E	F	F

The three strings are interconvertible through a series of single step rearrangements called inversions by geneticists, where the order of genes in a particular region of the chromosome becomes reversed by a mutational event. Thus sequence (1) may be converted into sequence (2) by inverting the order FE to EF, and so on. Although in such cases it is impossible to say which is the ancestral sequence, a number of single step conversions are possible:

$$2 \nearrow^{3}_{\searrow 1} \quad \text{or} \quad 1 \to 2 \to 3 \quad \text{or} \quad 3 \to 2 \to 1$$

The fact that the gene order in the chromosomes of the different species of fruit fly can be arranged in such perfect sequences is convincing evidence that species do originate in nature by descent from other species. If we want absolutely *bona fide* evidence for the reality of microevolutionary change and speciation in nature the cases of the circumpolar overlaps and the fruit fly of Hawaii come very close.

There are also now many examples of the second sort of evidence for speciation – theoretical reconstruction, step by step in great detail, of the evolutionary descent of one or more species from an ancestral species including a comprehensive and fully plausible account of precisely how each successive change occurred and hypothetical reconstruction of intermediate stages.

One classic study of this sort was Mengel's work on the wood warblers of North America.[7] At present there are something like forty-eight wood warblers in North America, differing in plumage and behaviour and adapted for environments as diverse as the semi-tropical forest of the south to the cool temperate forests of the north. Mengel was able to reconstruct in great detail a hypothetical genealogy of the group as they diversified from a single ancestral species into the many modern species over the past one million years. As well as a detailed geneology, he was also able to provide ecological and other explanations for all the major adaptive changes undergone along each phyletic line, as well as plausible reconstructions of the behaviour patterns and plumage of many of the ancestral and now extinct species and subspecies.

A similiar study was carried out by Amadon on the Hawaiian honeycreepers.[8] These are a group of birds unique to Hawaii which have diversified into twenty-two distinct species and forty-five subspecies from only one ancestral form. The diversity of their general morphology, beak morphology and behaviour is astonishing. This group includes one subspecies which has evolved a unique adaptation: the lower bill has become straight and heavy and is used in woodpecker fashion like a chisel to bore into the wood, while the upper mandible has become long and curved and is used as a probe to prize out insects revealed by the boring action of the lower bill. Like Mengel, Amadon was able to provide a detailed reconstruction of the entire history and genealogy of the group, showing in detail their descent from a common ancestral species.

Yet another analagous study was that carried out by Gorman and Atkins on a group of small lizards found on the arc of Caribbean islands which curve south from Puerto Rico to Grenada just off the South American coast.[9] Their analysis, based on chromosomal and enzyme patterns, enabled them to reconstruct the exact sequence of speciation events which had occurred as the lizards gradually colonized the islands.

In addition to having witnessed natural selection in operation and having securely documented the reality of speciation, modern evolutionary biology has also been able to provide a thoroughly worked out model showing, step by step, precisely how species formation occurs in nature.

The modern theory of speciation is based on knowledge drawn from a number of disciplines such as genetics, and particularly the genetics of populations which were unknown in the mid-nineteenth century. The most comprehensive and scholarly account is given by Mayr in his book *Animal Species and Evolution*.[10] According to the modern view, which is almost universally accepted nowadays, the geographical isolation of a population is the key event on the road to species formation. Geographical isolation prevents interbreeding with the parent population and allows the isolated daughter population to undergo unique adaptive changes leading eventually to the formation of a distinct subspecies. Later, when the newly evolved subspecies with its unique behavioural and adaptive characteristics comes into contact with its parent species, hybrids are at a selective disadvantage and isolating mechanisms evolve to prevent interbreeding and to preserve distinct and selectively advantageous adaptive features. Eventually full reproductive isolation evolves, converting the subspecies into a new species.*

Although this may not be the only mode of speciation – there is evidence that some species, especially plant species, have been formed suddenly by massive chromosomal mutations which create instantly a new daughter population reproductively isolated from its parental stock[11] – the geographical model advocated by Mayr and other leading authorities would seem to be the usual mechanism which applies to the great majority of animal and plant species.

The geographical model has been confirmed by extensive field studies and it is now well established that nearly every species is divided into a number of relatively isolated populations; and every taxonomic grade from race, subspecies, sibling species to true species is found in nature and every stage of the process of speciation from the differentiation of isolated populations to selection for hybrid sterility has been observed in innumerable cases. Although the human species has not diversified as dramatically as some other species it has been, nonetheless, subject to the same processes of change and differentiation into geographically isolated races which, under the influence of different selective pressures, have evolved unique adaptive characteristics such as the black skin of the negro, the white skin of the caucasian and, perhaps the most specialized of all racial adaptations, the upper lid of the mongoloid eye which may have evolved in response to the intense cold in central Asia in glacial times.

It is clear, then, that Darwin's special theory was largely correct. Natural selection has been directly observed and there can be no question now that new species do originate in nature; furthermore, it is now possible to explain in great detail the exact sequence of events that lead to species formation. Moreover, although there are some areas of disagreement among students of evolution as to the relative significance of natural selection as opposed to purely random processes

---

* The role of isolation was discussed by Darwin in Chapter Four of the *Origin*. He considered it important, but not an essential element in the process of speciation.

such as genetic drift in the process of speciation, no one doubts that natural selection plays an important role in the process. The validation of Darwin's special theory, which has been one of the major achievements of twentieth-century biology, has inevitably had the effect of enormously enhancing the credibility of his general theory of evolution.

The fact that organisms can undergo a considerable degree of evolution under perfectly natural conditions has always been one of the most persuasive facts conducive to an overall evolutionary view of nature. The observation that organisms had undoubtedly undergone drastic morphological differentiation in isolated regions such as the Galapagos Archipelago was seminal to the whole development of Darwin's own evolutionary position. In his *Evolutionary Notebook*, started in 1837, he wrote:[12]

> In July opened first note-book on "Transmutation of Species". Had been greatly struck from about month of previous March on character of S. American fossils – & species on Galapagos Archipelago. These facts origin (especially latter) of all my views.

For Darwin, all evolution was merely an extension of microevolutionary processes. Yet, despite the success of his special theory, despite the reality of microevolution, not all biologists have shared Darwin's confidence and accepted that the major divisions in nature could have been crossed by the same simple sorts of processes. Scepticism as to the validity of the extrapolation has been generally more marked on the European continent than in the English speaking world. The German zoologist, Bernhard Rensch,[13] was able to provide a long list of leading authorities who have been inclined to the view that macroevolution cannot be explained in terms of microevolutionary processes, or any other currently known mechanisms. These dissenters cannot be dismissed as cranks, creationists, or vitalists, for among their ranks are many first rate biologists. This is acknowledged by Mayr:[14]

> The nature and cause of transpecific evolution has been a highly controversial subject during the first half of this century. The proponents of the synthetic theory maintain that all evolution is due to the accumulation of small genetic changes, guided by natural selection, and that transpecific evolution is nothing but an extrapolation and magnification of the events that take place within populations and

species. A well-informed minority, however, including such outstanding authorities as the geneticist Goldschmidt, the paleontologist Schindewolf, and the zoologists Jeannel, Cuenot, and Cannon, maintained until the 1950's that neither evolution within species nor geographic speciation could explain the phenomena of "macroevolution", or, as it is better called, transpecific evolution. These authors contended that the origin of new "types" and of new organs could not be explained by the known facts of genetics and systematics.

However attractive the extrapolation, it does not necessarily follow that, because a certain degree of evolution has been shown to occur, therefore any degree of evolution is possible. There is obviously an enormous difference between the evolution of a colour change in a moth's wing and the evolution of an organ like the human brain, and the differences among the fruit flies of Hawaii, for example, are utterly trivial compared with the differences between a mouse and an elephant, or an octopus and a bee. While, admittedly, the adaptive radiations such as have occurred on oceanic archipelagos seem remarkably analagous to the great adaptive radiations of the major groups such as the mammals and the dinosaurs, there is an enormous difference in scale.

Whatever the merits of the extrapolation may be in biology, there are certainly many instances outside biology where such an extrapolation is clearly invalid, where large scale "macro" changes can only be accounted for by invoking radically different sorts of processes from those responsible for more limited "micro" types of change.

The sorts of phenomena such as the movement of high and low pressure systems which account for the "micro" day-to-day changes in the weather are quite incapable of explaining longer term "macro" changes such as the cycle of the seasons. This long-range periodic climatic pattern can only be explained by invoking astronomical factors, such as the tilt of the Earth's axis which shifts successively the northern, then the southern, hemisphere towards the sun as the Earth orbits the sun each year.

Geology offers further examples. Geological processes such as wind, water and ice erosion, the laying down of sedimentary rock, the periodic sinking and uplift of the Earth's crust, metamorphosis and volcanism, are perfectly capable of explaining much of the geomorphology of the earth's surface. But there are many more "macro" geological phenomena, such as the wandering of the poles revealed by recent paleomagnetic studies, the close morphologic fit between Africa and South

America and the similarity in the plant and animal species in the now widely scattered continents of the southern hemisphere, which can only be explained by the new theory of plate tectonics and its radical consequence that the continents are being continually moved across the surface of the planet, driven by massive convection currents welling up from deep inside the earth.

There are other lines of non-biological evidence which also seem to cast considerable doubt on the validity of the extrapolation. The technique of problem solving by trial and error, for example, a mechanism which is strictly analagous to evolution by natural selection, is often successful in solving relatively simple problems, but it would obviously be wrong to conclude that it is capable, at least in finite time, of solving more involved complex sorts of problems. Further, in the case of many complex non-living systems, although a certain degree of functional change can sometimes be achieved by a succession of minor functional modifications, there is invariably a limit to the degree of change which can be brought about in this way.

The example of language is instructive. A sentence, like an organism, is an example of a complex system. Each consists of a set of interacting components which are integrated together to form a meaningful proposition. Although it is possible to make minor changes in a sentence, Darwinian fashion, selecting successive single letter changes while at the same time avoiding ungrammatical or nonsense strings, there is a limit to the degree of change that can be achieved in this way. Consider the simple sentence:

He sat on the mat.

If we attempt to reach other meaningful sentences by changing one letter at a time and avoiding nonsensical sequences as we go, we soon discover that only a very small number of other sentences can be reached, such as:

She sat on the can.
We sat on the tram.

Hence we find that there exists a small group of sentences related to one another and interconvertible by unit mutational steps but isolated from all other groups of sentences by the severity of the rules of English grammar. It is impossible to reach even a closely related sentence such as

He stood on the mat.

in unit mutational steps without passing through nonsensical strings:

He sat on the mat.
He stt on the mat.
He sto on the mat.
He stoo on the mat.
He stood on the mat.

It is only possible to go from one isolated island of sentences to another by making two or more changes simultaneously – a sort of macromutational reorganisation of the letter sequence – thereby jumping over the intervening nonsense barrier to another island of meaning. For example, if we make four simultaneous changes to the sentence:

He sat on the mat.

we can get to   He stood on the mat.   and from this new sentence a further set of related sequences are obtainable by unit mutational changes.

More complicated sentences are even more isolated. For example, the sentence:

Most English sentences are complex sequences of letters.

is so isolated that no other grammatical string of letters can be reached by making single letter substitutions.

In this case, to reach new sentences requires, at a minimum, single word substitutions such as those shown below:

most → all → some   complex → complicated → long
English → French   sequences → strings → arrangements
sentences → words   of
          are   letters → words

But again, as in the case of much simpler sentences, it is impossible to reach through a succession of single word substitutions any more than a handful of related sentences.

most	English	sentences	are	complex	sequences	of	letters
some	,,	,,	,,	,,	,,	,,	,,
,,	French	,,	,,	,,	,,	,,	,,
,,	,,	,,	,,	long	,,	,,	,,
,,	,,	words	,,	,,	,,	,,	,,

To get to more distantly related sentences we must make two or more single word substitutions simultaneously, that is, a radical macro-mutational change in the word sequence.

most   English   cities   are   complex   arrangements   of   roads

It is evident that sentences are separated from one another by distinct discontinuities which cannot be crossed gradually through a succession of minor changes. Sentences, in fact, cannot undergo "Darwinian evolution" except to a very limited degree.

As a general rule, the longer the sentence and the more complex the information it conveys, the more isolated it will be from all other sentences in the language and the less amenable it will be to any sort of gradual functional transformation. Paragraphs consisting of a set of related sentences are even more resistant to any sort of gradual functional change. To change significantly the meaning of a paragraph requires a gigantic "macromutational" event necessitating so many simultaneous changes in the word sequence that, in effect, we have to rewrite the paragraph. What holds for natural languages also, of course, holds for artificial computer languages where, again, "sentences" cannot be transformed gradually Darwinian fashion through a series of unit mutational steps.

The same rule applies in the case of most other sorts of complex systems where function arises from the integrated activity of a number of coadapted components. Take the case of a watch, where only very trivial changes in the structure and function of the cogwheel system can be achieved gradually through a succession of minor modifications. Any major functional innovation, such as the addition of a new cogwheel or an increase in the diameter of an existing cogwheel, necessarily involves simultaneous highly specific correlated changes throughout the entire cogwheel system. Like a sentence, the function of a watch cannot be gradually converted through an innumerable series of transitional forms into a quite different sort of watch. It is the same with automobile engines. If the diameter of the cylinder is increased, function can only be conserved if there are compensatory changes in the diameter of the piston, strength of the connecting rods and changes in the design of the cylinder head.

While sentences, machines and other sorts of complex systems can undergo a certain degree of gradual functional change, there is invariably a limit beyond which the system cannot undergo further gradual change. To cross as it were from one "type" of system to another necessitates a relatively massive reorganization involving the redesign or respecification of all or most of the interacting component subsystems. Systems can undergo gradual microevolution through a succession of minor changes in their component structures but macroevolution invariably involves a sudden "saltational" change. Clearly, in all such cases, the extrapolation from micro- to macroevolutionary change does not hold.*

Living organisms are certainly analogous to other complex systems in many ways. Might it not be true for living systems, as it is for systems such as sentences or watches, that function is distributed according to a discontinuous pattern and that the gaps between the islands of function are unlinked by functional transitional systems? The fact, already referred to in Chapter Two, that the degree of change that can be experimentally induced in a wide variety of organisms, from bacteria to mammals, even under the most intensive selection pressures, is always limited by a distinct barrier beyond which further change is impossible would seem to bear out the analogy.

It was Cuvier's view that living organisms could only undergo a limited degree of change, and for precisely those reasons which apply in the case of any other complex system – that each type is a unique adapted whole, its parts perfectly fashioned to undergo coherent interactions. As he wrote:[15]

> Every organized being forms a whole, a unique, and perfect system, the parts of which mutually correspond and concur in the same definitive action by a reciprical reaction. None of these parts can change without the whole changing.

From which it followed, just as in the case of any sort of functioning machine, that because fundamental change in the design of an organism would require simultaneous and coadaptive change throughout the organism to ensure that all the components would still function together in a coherent and integrated manner and, because such a sudden adaptive reorganization involving all the parts of an organism seemed to Cuvier and Agassiz and indeed all early nineteenth-century biologists quite literally inconceivable, this completely precluded any sort of gradual transformation.

* The reason why complex systems cannot undergo gradual functional transformation via a succession of small changes is discussed in Chapter 13.

There is no doubt that the success of the Darwinian model in explaining microevolution invites the hope that it might be applicable also to macroevolutionary phenomena. Perhaps in the end this might prove to be the case; but, on the other hand, there is the depressing precedent, as the history of science testifies, that over and over again theories which were thought to be generally valid at the time proved eventually to be valid only in a restricted sphere. Newtonian physics, for example, which accounted perfectly for all the empirical data available in the eighteenth and nineteenth centuries and is still used for calculating the trajectory of a space rocket, is absolutely inapplicable to phenomena at the subatomic and cosmological levels. Theories are seldom infinitely extendible.

NOTES

1. Diver, C. (1936) "The Present State of the Theory of Natural Selection", special symposium, *Proc. Roy. Soc. B*, 121: 43–73, see p62.
2. Kettlewell, H. B. D. (1959) "Darwin's Missing Evidence", *Scientific American*, 201 (3); 148–53, see p49.
3. Mayr, E. (1970) *Populations, Species and Evolution*, Harvard University Press, Cambridge, Mass, p279.
4. ibid, p291.
5. ibid, p290.
6. Dobzhansky, T., Ayala, F. J., Stebbins, G. L. and Valentine, J. W. (1977) *Evolution*, W. H. Freeman and Co, San Francisco, pp271–76.
7. Mengel, R. M. (1964) "The Probable History of Species Formation in some Northern Wood Warblers (*Parulidae*)", *The Living Bird*, 3: 9–43.
8. Amadon, D. (1950) "The Hawaiian Honeycreepers (*Aves Drepaniidae*)", *Bull. Amer. Mus. Nat. Hist.*, 95: 157–262.
9. Gorman, G. C. and Atkins, L. (1969) "The Zoogeography of Lesser Antillean Anolis Lizards", *Bull. Mus. Comp. Zoology*, 138: 53–80.
10. Mayr, E. (1963) *Animal Species and Evolution*, Harvard University Press, Cambridge, Mass.
11. White, M. J. D. (1969) "Chromosomal Rearrangements and Speciation in Animals", *Ann. Rev. Genet.*, 3: 75–98.
12. Darwin, F. (1882) *The Life and Letters of Charles Darwin*, 3 vols, John Murray, London, vol 1, p276.
13. Rensch, B. (1959) *Evolution above the Species Level*, Columbia University Press, New York, p57.
14. Mayr. (1970), op cit, p351.
15. Cuvier, G. (1829) *Revolutions of the Surface of the Globe*, English edition, Whittaker, Treacher and Arnot, London, p60.

CHAPTER 5
The Typological Perception of Nature

"Natura non facit saltum."
This canon if we look to the present inhabitants of the world is not strictly speaking correct.

Evidence for evolution exists in nature wherever a group of organisms can be arranged into a lineal or sequential pattern, in which case the idea of evolution becomes almost irresistible. The circumpolar overlaps and the chromosomal pattern of the fruit flies of Hawaii are compelling evidence for evolution because in each case the sequential arrangement corresponds to an almost perfect continuum.

Obviously, the more perfect the sequence the more convincing it is as evidence for evolution, but even when there is no perfect continuum of forms, just as long as there is a clear sequential order to the patterns of diversity, the conclusion of evolution is still very difficult to resist. A classic case is the series of fossil horses (see Chapter Eight). This series is nothing like a perfect continuum of forms, the breaks are distinct and clear, but the overall sequential pattern is so obvious that no one seriously doubts that the modern horse has evolved from the primitive horses of the Eocene era sixty million years ago.

No one would doubt that whales had evolved gradually from an ancestral land mammal if there was a complete sequence of forms leading gradually from a small otter-like species through seal-like organisms to the whales. Would anyone doubt the Darwinian claim that all mammals have evolved from a common ancestral species if all known mammalian species, including such diverse forms as whales and bats, could be connected together through a series of transitional forms leading ultimately back to one original source?

It was primarily because most groups of organisms seemed so isolated and unlinked by transitional forms that for the better part of a century, prior to 1859, most biologists saw the facts of biology as pointing to a model of nature which was diametrically opposed to, and indeed irreconcilable, with the notion of organic evolution. Something of the diametric opposition many nineteenth-century biologists felt towards the idea of evolution is evidenced in the preface of Louis Agassiz's book *Methods of Study in Natural History*, published in 1863 just four years after the publication of the *Origin*. For Agassiz it was only the inherent intellectual appeal of the concept of continuity which inclined naturalists to interpret nature in evolutionary terms. To seek continuity in the pattern of nature was futile:[1]

> . . . It is my belief that naturalists are chasing a phantom, in their search after some material gradation among created beings, by which the whole Animal Kingdom may have been derived by successive development from a single germ, or from a few germs. It would seem, from the frequency with which this notion is revived, – ever returning upon us with hydra-headed tenacity of life, and presenting itself under a new form as soon as the preceding one has been exploded and set aside, – that it has a certain fascination for the human mind.

Agassiz, in common with the great majority of leading biologists in the nineteenth century, adhered to a philosophy of nature referred to as typological which was completely antithetical to the concept of organic evolution and which denied absolutely the existence of any sort of sequential order to the pattern of nature.

According to the typological model of nature all the variation exhibited by the individual members of a particular class was merely variation on an underlying theme or design which was fundamentally invariant and immutable. Each individual member of a class conformed absolutely in all essential details to the theme or archetype* of its class. It followed from this that all the members of a class were equally representative and characteristic of their class and that no individual member could be considered in any fundamental sense any less characteristic of its class or any closer to any member of another class than any of the other members of its class. In other words, all the members of any defined class are equidistant from the

---

*In typological theory the term archetype referred to a purely hypothetical entity, an abstract and highly generalized representative of its class.

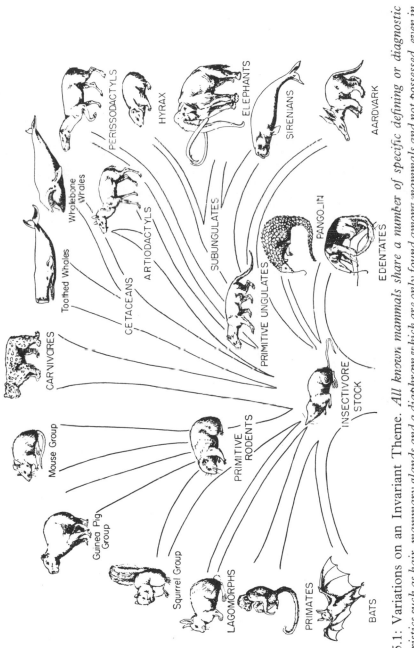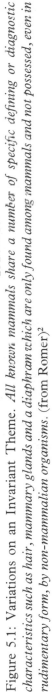

Figure 5.1: Variations on an Invariant Theme. *All known mammals share a number of specific defining or diagnostic characteristics such as hair, mammary glands and a diaphram which are only found among mammals and not possessed, even in rudimentary form, by non-mammalian organisms.* (from Romer)[2]

members of other classes as well as being equirepresentative of the archetype of their class.

Such a model of biological classes completely excluded any sort of significant sequential order to the pattern of nature. Because no member of any defined class could stray beyond the confines of its type in terms of its basic characteristics, then no class could be led up to gradually or linked to another class through a sequence of intermediates. Further, within one class, because all the members conform absolutely to the same underlying design and are equidistant in terms of their fundamental characteristics from all other classes, it is impossible to arrange them into a sequence leading in any significant sense towards another class. Typology implied that intermediates were impossible, that there were complete discontinuities between each type.

According to typology, biological classes were analogous to classes of geometrical figures. Note in the diagram below how *all* the members of each class are equally characteristic of their class. Each triangle (class A) exhibits to full degree the fundamental defining characteristic of its class, ie each is a figure bounded by three straight lines. Similarly, each quadrilateral is equally characteristic of its class, ie each is a figure bounded by four straight lines.

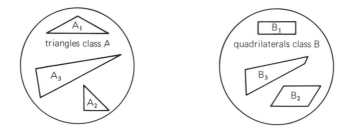

Because the members of each class of figures are perfectly characteristic of their class then the divisions between the two classes are absolute and unbridged by "intermediate" forms or sequences leading from one class to another. When the members of one class are compared with the members of another, all are found to be equally isolated in terms of their fundamental characteristics. For example, all triangles are equidistant from any quadrilateral.

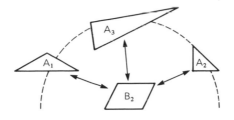

Where the representative of one class happens to resemble the representative of another class the resemblance is only superficial and not indicative of any sort of profound relationship. For example, triangle A3 and quadrilateral B3 resemble each other, but both are still perfectly representative of their respective classes.

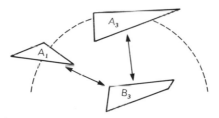

Translated back into biological terms, typology implied that all mammalian species, for example, were equirepresentative of the mammalian archetype:

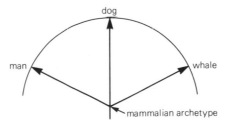

and all species of birds were equirepresentative of the avian archetype:

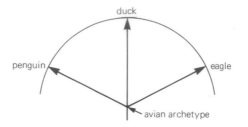

When the species of one class, say birds, were compared with any non-avian species all were equidistant in terms of their fundamental avian characteristics so that no species of bird was fundamentally closer to any non-avian species.

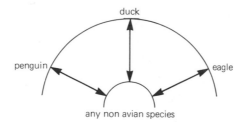

Typology implied that there were absolute discontinuities between each class of organisms, that life was therefore fundamentally a discontinuous phenomenon and that sequential arrangements, whereby different classes were linked together or approached gradually through series of transitional forms, should be completely absent from the entire realm of nature. Typology contrasted completely with the idea of organic evolution. Mayr emphasizes the opposition of the two antithetical views of nature:[3]

> The concepts of unchanging essences and of complete discontinuities between every eidos (type) and all others make genuine evolutionary thinking well-nigh impossible. I agree with those who claim that the essentialist philosophies of Plato and Aristotle are incompatible with evolutionary thinking. . . . The assumptions of population thinking (evolution) are diametrically opposed to those of the typologists. . . . The ultimate conclusions of the population thinkers (evolutionists) and of the typologist are precisely the opposite. For the typologist, the type (eidos) is real and the variation an illusion, while for the popu-

lationists (evolutionists) the type (average) is an abstraction and only the variation is real. No two ways of looking at nature could be more different.

Typology acknowledged the existence of biological variation but denied that it could ever be radical or directional. It was fundamental to typology that variation was always conservative and limited, always intratype, and never inter-type as is required by evolution.

Nearly all the great biologists and naturalists of the late eighteenth and early nineteenth centuries who founded the modern disciplines of comparative anatomy, taxonomy and paleontology adhered strictly to a discontinuous typological model of nature. This constellation included the Swedish botanist Carl Linnaeus who in 1735 founded modern taxonomy with his *Systema Naturae*. Another leading typologist was the French biologist Georges Cuvier, who virtually founded vertebrate paleontology and comparative anatomy and was well known in his day because of his use of his principle of correlation to reconstruct the morphology of entire vertebrate species from single bones. He was the first to establish clearly that fossils were the remains of now extinct forms, different from but allied to living creatures, and was also the first to suggest that fossils could be used to establish the age of geological strata. And, of course, there was Louis Agassiz, who is perhaps largely remembered for advancing the theory that vast areas of the earth had once been covered by ice.

Then there was Richard Owen, probably the leading nineteenth-century British anatomist, who produced a classification system for fossil reptiles and originated the term dinosaur. He was also the first director of the Natural History Museum at South Kensington in London. Another member of this illustrious group of naturalists was the Scottish geologist Charles Lyell, who was for most of his life vigorously opposed to the idea of evolution. He won universal fame for his *Principles of Geology* which ran to eleven editions in his lifetime and had a decisive influence on the future development of the science. His main thesis was the uniformity of natural forces in the past and present working in a process of slow and unending change, which has since become a basic tenet of geology. He first suggested the terms Eocene, Miocene, Pliocene and Pleistocene for the divisions of the Tertiary and attempted to define the limits of the Cambrian and Silurian.

The fact that so many of the founders of modern biology, those who discovered all the basic facts of comparative morphology upon which modern evolutionary biology is based, held nature to be fundamentally a discontinuum of isolated and unique types unbridged by transitional varieties, a position absolutely at odds with evolutionary ideas, is obviously very difficult to reconcile with the popular notion that all the facts of biology irrefutably support an evolutionary interpretation.

One of the traditional escapes from this dilemma is to presume that these scientists' typological model of nature was derived not from the facts of nature but from religious and metaphysical preconceptions which were prevalent at that time. This particular rationalization was advocated by many leading evolutionists shortly after 1859, including even the moderates like Joseph Hooker,[4] and soon became widely accepted. It has persisted as one of the great myths of twentieth-century biology, with the unfortunate consequence that the views of the pre-evolutionary biologists are nowadays largely ignored or considered archaic and overly influenced by religious belief.

However, it is quite likely that their religious prejudices have been exaggerated. Lyell, for example, firmly believed the earth to have been far older that the six thousand years implied in Genesis and about Cuvier, Coleman has written:[5]

> He was not, however, a doctrinaire Biblical zoologist. As a basis for intelligent discussion of the problems of natural history he preferred nature and the animals to ancient nonzoological authority. Cuvier saw presented in Scripture instructions for increasing the happiness and moral well-being of mankind and not texts for the exact study of natural history; this study was the province of science and not of theology.

Coleman is not the only authority who has cautioned against dismissing the anti-evolutionary stance of nineteenth-century biology as religious prejudice. Martin Rudwick, for example, also argues that the opposition of nineteenth-century biology to the ideas of evolution was not primarily theological. Richard Owen seems to have been perfectly prepared to accept some sort of secondary or natural cause for the progression of life on earth:[6]

> This is not to say that there were no metaphysical or theological components in the opposition of scientists such as Richard Owen, who

wrote one of the first and most important critical reviews of the *Origin* (and who is said to have coached Wilberforce for his British Association speech). But Owen's criticism of Darwin's theory was not a defence of biblical literalism or of special creation.

As Rudwick stresses, Owen could see no evidence among living organisms or in the fossil record for the idea of gradual transform- ations, which was the crucial point of departure for him and many other opponents of evolution before 1859. His dissent was empirical, not theological. As Hull cautions:[7]

> Too often, all the opponents of evolutionary theory are lumped together and their persistence explained away as religious bigotry.

There is, of course, ultimately no reason why theology should demand discontinuity of nature. Ironically, in the seventeenth and eighteenth centuries, many biologists and philosophers, influenced by the doctrine of the plentitude of creation and its corollary, the concept of the great chain of being, saw in theology a demand for continuity just as absolute as that demanded by modern evolutionary biology.

There is no doubt that the leading advocates of typology would have strenuously denied the idea that their views were largely derived from metaphysics or religion. On the contrary, they clearly believed them to be grounded in empiricism and observation. For Agassiz it was the evolutionists, not the typologists, whose views were pre- judiced by *a priori* concepts, and who were "chasing a phantom". When Cuvier expressed puzzlement at the reasons for the constancy of certain apparently non-adaptive defining characteristics within certain classes, he did not turn to religion or metaphysics:[8]

> Whatever secret reasonings there may be in these constant relations, it is *observation* which has elicited them, independently of general philosophy.
>
> *[emphasis added]*

When he argued for the constancy of the type he always turned to the facts of nature, never to metaphysics. The fossils, for example, provided no empirical evidence for change.[9]

> . . . if the species has gradually changed, we must find traces of these gradual modifications; that between the palaeotheria and the present

species we should have discovered some intermediate formation; but to the present time none of these have appeared. Why have not the bowels of the earth preserved the monuments of so remarkable a genealogy, unless it be that the species of former ages were as constant as our own . . .

The fact that it was possible to predict the entire morphology of an organism from only a tiny fragment of one of its parts provided Cuvier with what he saw as irrefutable evidence for the fundamental invariance of particular types and of all their unique defining characteristics.

The typological view of the immutability of the basic types of nature was not only based on the empirical observation that there was a fundamental constancy to the characteristics of each type but was also based on consideration of the fact that each different kind of organism was a uniquely adapted whole. Considering the limbs of a carnivore Cuvier wrote:[10]

That the claws may seize the prey, they must have a certain mobility in the talons, a certain strength in the nails, whence will result determinate formations in all the claws, and the necessary distribution of muscles and tendons; it will be necessary that the fore-arm have a certain facility of turning, whence again will result determinate formation in the bones which compose it; but the bone of the fore-arm, articulating in the shoulder-bone, cannot change its structure, without this latter also changes. The shoulder-blade will have a certain degree of strength in those animals which employ their legs to seize with, and they will thence obtain peculiar structure. The play of all these parts will require certain properties in all the muscles, and the impression of these muscles so proportioned will more fully determine the structure of the bones.

Because all the parts of each organism were so beautifully fashioned to function together it seemed self-evident to Cuvier that any major functional transformation would necessitate simultaneous coherent coadaptive changes in all of its component structures, but as such a sudden purposeful reorganization of all the component structures of an organism was so vastly improbable this seemed to preclude any sort of evolutionary transmutation. As Rudwick puts it for Cuvier and the other biologists of those times:[11]

To believe that such intricately coordinated organic mechanisms had come into being by "chance" or "accident", as theories such as Geoffroy's were felt to imply, was literally inconceivable.

The idea of transmutation was rendered even less likely in the eyes of many nineteenth-century typologists by the well-established fact that breeding experiments with domestic animals had for generations revealed a distinct limit beyond which further change became impossible.

Thus for Cuvier typology was grounded in empiricism and rationalism. If it had any relationship to religion or metaphysics its connection, as anyone who has read him will testify, was only the most tenuous. Similarly, for Owen, there was simply no empirical evidence for believing that the sort of gradual evolution by natural selection conceived of by Darwin had ever in fact occurred:[12]

> Is there any one instance proved by observed facts of such transmutation? When we see the intervals that divide most species from their nearest congeners, in the recent and especially the fossil series, we either doubt the fact of progressive conversion, or, as Mr Darwin remarks in his letter to Dr. Asa Gray, one's 'imagination must fill up very wide blanks.'
> The last ichthyosaurus, by which the genus disappears in chalk, is hardly distinguishable specifically from the first ichthyosaurus, which abruptly introduces that strange form of sea-lizard in the Lias. The oldest Pterodactyle is as thorough and complete a one as the latest.

Likewise, one of the leading nineteenth-century German palaeontologists, H. G. Bronn, was unable to accept Darwin's views because, as he wrote in a review of *The Origin of Species*, he found[13] "the enormous gaps which now confront us in the series of plant and animal forms" impeded his "complete consent".

Another leading continental paleontologist was François Jules Pictet. Although he found Darwin's arguments quite persuasive:[14]

> . . . his theory squares quite well with the great facts of comparative anatomy and zoology. It corresponds equally well with many palaeontological facts. It accords well with the specific resemblances which exist between two consecutive faunas

but again the crucial empirical evidence of intermediates was absent:[15]

> Why don't we find these gradations in the fossil record, and why, instead of collecting thousands of identical individuals, do we not find more intermediary forms? To this Mr. Darwin replies that we have only a few incomplete pages in the great book of nature and the transitions have been in the pages which we lack. By why then and by what peculiar rules of probability does it happen that the species which we find most frequently and most abundantly in all the newly discovered beds are in the immense majority of the cases species which we already have in our collections?

He was forced to conclude:[16]

> Thus we find ourselves in a singular position. We are presented with a theory which on the one hand seems to be impossible because it is inconsistent with the observed facts and on the other hand appears to be the best explanation of how organized beings have been developed in the epochs previous to ours.

Anyone prepared to read the views of the leading opponents of evolution in the nineteenth century (a convenient source is a recently published collection of reviews edited by David Hull) will be forced to conclude that it was the absence of factual evidence which was the primary source of their scepticism and not religious prejudice. Gould had even gone as far as claiming:[17]

> Contrary to popular belief, no serious nineteenth-century scientist – not even the most theological catastrophist – argued for the direct intervention of God in the earth's affairs. All accepted the constancy of natural law. God . . . did not need to meddle by miracle with the subsequent history of the earth.

Although Gould is clearly exaggerating (after all, Darwin himself is quite insistent [see discussion at the end of the concluding chapter of the *Origin*] that most of his opponents accepted the idea of the independent origin of each species by "miraculous acts of creation", and it is difficult to see what a miraculous act of creation can mean if it does not involve God meddling in nature), nonetheless there is an element of truth in Gould's remarks. The fact is that the anti-evolutionism of most leading nineteenth-century biologists was not primarily religious. They did not leap to the creationist model. If they turned to miracle it was because in the final analysis they could see no conceivable naturalistic alternative.

Even if we allow that the typological views if the early nineteenth-century biologists were influenced to some extent by religion or metaphysics it can hardly be denied that there has always been massive empirical evidence for the typological model of nature within the existing realm of life. Admittedly, the axioms of typology have been shown to be inapplicable at the level of the species. Species can and do evolve and many can be linked to other species through clear sequences of intermediate subspecies; consequently, distinct demarcations cannot be drawn at the lowest taxonomic levels. But, at levels above the species, the typological model holds almost universally. Indeed, the isolation and distinctness of different types of organisms and the existence of clear discontinuities in nature have been self-evident for centuries, even to non-biologists. No one, for example, has any difficulty in recognizing a bird, whether it is an eagle, an ostrich or a penguin; or a cat, whether it is a domestic cat, a lynx or a tiger. Moreover, no one can name a bird or a cat which is in any sense not fully characteristic of its class. No bird is any less a bird than any other bird, nor is any cat any less a cat or any closer to a non-cat species than any other cat.

The reason for the distinctness of each class and the absence of sequential arrangements, whereby classes can be approached gradually through a series of transitional forms, is precisely as typology implied because each class of organism (just like a class of geometric figures) possesses a number of unique defining characteristics which occur in fundamentally invariant form in all the species of that class but which are not found even in a rudimentary form in any species outside that class. Take, for example, mammals. All the members of this class exhibit a number of unique features which are not found in any other group of organisms. They include: a hairy integument, each hair being a complex structure consisting of a keratinized cuticle, a cortex and a central medulla; mammary glands exhibiting alveoli surrounded by a network of myoepithelial cells responsive to the hormone oxytocin producing milk, a nutritious secretion containing fat globules and sugars; specialized sweat glands in the skin; a four-chambered heart with left ventricle delivering aereated blood to the aorta; discrete and reniform kidneys, with nephron form and function specialized to generate a concentrated urine containing a high concentration of urea; a large cerebral cortex with distinctive six layers of cells; a diaphragm, a special muscle used by mammals for respiration; three highly specialized ear ossicles – a mallus, incus and stapes conducting vibrations across the middle ear; the organ of corti, a specialized organ for reception and analysis of sound.

Each of these characteristics are exhibited by *all* mammals in essentially invariant form. Although there is variation it is only trivial, variation on an invariant theme. Take hair, for example. A typical mammalian hair is a rather complex structure. It consists of an expanded root and a shaft below the skin in an epidermal sheath or hair follicle. Generally one or more sebaceous glands drain into the cleft between the hair shaft and adjacent tissues. The follicle slants at an angle to the skin surface and a tiny smooth muscle runs downward from the outer part of the dermis to insert near the base of the shaft. Hair is a diagnostic characteristic of mammals occurring in no other class of organism. Its form varies from the stiff quills of the porcupine to the soft wool hair which forms the major component of the pelt of many common species of mammals. But, despite these specializations, the basic design of mammalian hair is invariant: moreover, no structures are known which can be considered in any sense transitional between hair and any other vertebrate dermal structure.

Another diagnostic characteristic of mammals is the cerebral cortex. This is a complex outgrowth of neural tissue which forms the outer layer of the brain and which is the seat of all the higher mental functions and complex behaviour patterns so characteristic of mammals. It consists of millions of nerve cells and gives rise to a number of neural pathways which are quite unique to the mammalian brain. In all mammals the nerve cells in the cortex are organized into the same invariant pattern of six basic layers of cells and the same regions are devoted to the analysis of the same sorts of information, visual, motor or sensory. Although the cortex varies enormously in different mammalian groups, being highly developed in primates and the Cetacea (the whales and dolphins), but only poorly developed in the insectivores (the shrews and hedgehogs), its basic design is fundamentally invariant. There is nothing like it in the brain of any non-mammalian vertebrate and it is not led up to gradually through a sequence of less complex neurological structures in any known group of organisms.

Similarly, birds possess a number of diagnostic characteristics which are absolutely unique, including, for example: the feather; a unique arrangement of flight feathers on their wings which form the basis of an aerofoil; a unique continuous flow-through lung system; vastly enlarged cerebral hemispheres of completely different structure to that of a mammal.

A similar suit of unique defining characteristics could be assembled for a host of other biological classes – from insects to flowering plants – and because in every case all the members of each class possess to a similar degree all the defining characteristics of their class, then in terms of these defining characteristics no members of the class is any less characteristic of its class than any other.

If one were to list all identifiable groups of organisms currently in existence which perfectly satisfy the axioms of typology we would end up naming virtually every single identifiable taxon in nature. And as far as the individual defining characteristics are concerned, one could continue citing almost *ad infinitum* complex defining characteristics of particular classes or organisms which are without analogy or precedent in any other part of the living kingdom and are not led up to in any way through a series of transitional structures. Such a list would include structures as diverse as the vertebral column of vertebrates, the jumping apparatus of the click beetle, the pentadactyl limb of tetrapods, the spinneret and male copulating organ of spiders, the wing of a bat, the water vascular and ambulacral systems of echinoderms, the neck of the giraffe, the male reproductive organs of the dragonfly, and so on until one had practically named every significant characteristic of every living thing on earth.

In addition to the character traits used in taxonomy, there are other character traits which occur widely and non-systematically in diverse groups of organisms and which cannot therefore be used for taxonomic purposes, but which nonetheless also exhibit an essentially invariant form in all the species in which they occur and are not led up to gradually through a sequence of intermediate structures. A fascinating example is the Cilium.

Cilia are tiny microscopic hairs which project from the surface of cells. They occur on a vast number of different cell types from the respiratory epithelium in mammals to the gill surfaces in molluscs. They occur in nearly all animal species as well as in some protozoans and plants. They have been known since the time of Leeuwenhoek in the seventeenth century but it was only recently, after their molecu-

lar structure was examined by electronmicroscopy, that their uniqueness and isolation from all other structures in the living world was
finally revealed.

The structure of cilia is described by Peter Satir in a *Scientific
American* article in 1974:[18]

> Each cilium could be seen to consist primarily of a sheaf of filaments,
> or tubular elements, arranged in what came to be called the $9 + 2$
> pattern, with nine doublet elements surrounding a central pair of
> singlets. The sheaf of filaments, called an axoneme, was surrounded
> by and enclosed in an extension of the cell membrane, which meant
> that it was essentially a protruding portion of the cytoplasm of the cell.

But the most remarkable feature of the organelle is described by
Frey-Wyssling:[19]

> One of the most fascinating discoveries of electron microscopy disclosed
> that all cilia, from the flagellates and ciliates to the lower plants with
> antherozoids and through all phyla of the animal kingdom up to the
> mammalian including human sperms and ciliated epithelia are of the
> $9 + 2$ stranded type . . . what we must recognise is an amazing
> conservatism which clings, so to speak, indefinitely to an established
> and approved structure.

Apart from variation in length the only other variation in structure
that has ever been observed is the absence of the two central filaments
in one or two cases, such as in the tail of certain spermatazoa. Every
cilium that has been examined to date has been found to possess
essentially the same basic structure. No cilia are known which possess,
for example, 3, 5 or 7 filaments or possess the filaments in any other
but the typical $9 + 2$ arrangement. There is no hint anywhere of any
sort of structure halfway to the complex molecular organization of
these fascinating microhairs through which their evolution might
have occurred.

The genetic code is perhaps the most fundamental of all the
adaptations of the living world. It is the crucial system upon which
the function of every living cell on earth depends. As it is the subject
of extensive discussion in later chapters, it will not be discussed in
detail here save to mention that, as a result of one of the most
remarkable discoveries in molecular biology, it is now known to be a
unique and invariant system of rules which is identical in every cell

on earth. No cell has ever been found that departs in any significant way from the universal pattern of the code. Apart from artificial language used in computers and human language itself, the genetic code, or the language of life as it has been called, is without any analogue in the physical universe. Like cilia and like so many of the characteristics found in living things on earth, the genetic code is not led up to gradually through a sequence of transitional forms.

It is a remarkable testimony to the almost perfect correspondence of the existing pattern of nature with the typological model that, out of all the millions of living species known to biology, only a handful can be considered to be in any sense intermediate between other well defined types.

The lungfish is a classic example. It has fins, gills and an intestine containing a spiral valve like any fish but lungs, heart and a larval stage like an amphibian. Another classic example of an intermediate type is the egg-laying mammals, the monotremes such as the duck billed platypus. In laying eggs the monotremes are reptilian, but in their possession of hair, mammary glands, and three ear ossicles they are entirely mammalian.

Undoubtedly, if the various anatomical and physiological systems in the lungfish and the monotremes were all strictly transitional between fish and amphibia and between reptiles and mammals respectively, then the case for them being genuine transitional types would be far clearer. However, in the case of lungfish, its fish characteristics such as its gills and its intestinal spiral valve are one hundred per cent typical of the condition found in many ordinary fish, while its heart and the way the blood is returned to the heart from the lungs is similar to the situation found in most terrestrial vertebrates. In other words, although the lungfish betrays a bewildering mixture of fish and amphibian character traits, the individual characteristics themselves are not in any realistic sense transitional between the two types.

The biology of the monotremes is similar. Again, where they are reptilian in, for example, the reproductive system and in the structure of their eggs, they seem almost fully reptilian, while where they are mammalian, as for example in the construction of their middle ear, or in the possession of hair, they are fully mammalian. Instead of finding character traits which are obviously transitional we find them to be either basically reptilian or basically mammalian, so that although the monotremes are a puzzle in terms of typology they afford little evidence for believing that any of the basic character traits of the mammals were achieved gradually in the way evolution envisages.

Yet another classic aberrant 'intermediate' is the small caterpillar-like organism *Peripatus*, a member of the phylum Onychophora, considered to be intermediate between the annelid worms and the arthropods. But once again, as in the case of the lungfish and the monotremes, its organ systems are not strictly transitional between the two groups. For example, the circulatory and respiratory systems of *Peripatus* are quite typically arthropod in their basic design, while its nervous and excretary systems are quite typical of those seen in many annelid worms. *Peripatus*, like the lungfish and the platypus, is really a mosaic of characteristics drawn from two distinct groups.

There is no question that those forms are somewhat anomalous in terms of typology, each exhibiting a curious combination of the diagnostic characteristics of two otherwise quite distinct types. But they provide little evidence for believing that one type of organism was ever gradually converted into another. They cannot be construed as evolutionary links except in the vaguest sort of way. Not only are their individual organ systems untransitional, but in many aspects of their biology all three types are extremely isolated from the group they supposedly unite. Between lungfish and amphibia, between monotremes and reptiles and between *Peripatus* and arthropods, there are tremendous gaps unbridged by any transitional forms.

The fact that the only living 'intermediates' known to biology are a handful of species such as the lungfish and the monotremes among the vertebrates, and the Onychophora among the invertebrates, organisms which undoubtedly exhibit an extraordinary mosaic-like combination of characteristics but which can only be construed as being intermediate in the vaguest sense of the word, merely under-lines the fact that on the whole most existing taxa are remarkably well defined and strikingly isolated. Not only are *bona fide* intermediates virtually unknown, it is impossible to allude to any more than a handful of cases where the pattern of nature seems to exhibit some-thing of sequential argument. One of the best known of such supposed sequential arrangements is the vertebrate series from the cyclostomes through fish, amphibia, and reptiles to mammals.

Every school text teaches the story of vertebrate evolution as a series of successive transformations from fish to man. According to the traditional view of vertebrate evolution, the vertebrates first originated about six hundred million years ago. The ancestral proto-vertebrates are considered to have been primitive, jawless, fish-like creatures similar to the living cyclostomes of today, such as the lamprey. Sometime, possibly in the late Cambrian age about five hundred million years ago, probably in fresh water, a certain group of these primitive jawless fishes supposedly gave rise to a number of higher classes of jawed fish. These included the archaic and now extinct heavily armoured Placoderms – the dominant fish of Devonian times – and the cartilaginous fish represented today by the sharks and rays, and the bony fish which are the dominant fish of today. The bony fishes were a diverse group. When they first appeared in the fossil record they were already well differentiated into a number of clearly defined classes. About four hundred million years ago one group of bony fish closely related to the present day lungfish of Africa gave rise to the amphibia which in turn gave rise to the reptiles. The reptiles diversified into many different forms including the now extinct dinosaurs, which dominated the earth for many millions of years right up to the age of mammals. Sometime around two hundred and fifty million years ago a group of reptiles known as mammal-like reptiles supposedly gave rise to the mammals which eventually diversified into the modern placental and marsupial groups.

The sequence is defended on the grounds that in terms of their basic morphology the groups seem to fall into a natural sequence. It is claimed that, based on the morphology of their brain and nervous system, heart and cardiovascular system, alimentary tract and skeletal system, amphibia, for example, can be considered to be intermediate between, on the one hand, bony fish and, on the other hand, reptiles. On the same morphological criteria it is claimed that the reptiles can be considered intermediate between the amphibia and the mammals. While there may be some justification for accepting this standard view which is presented in most textbooks – that the vertebrate classes from cyclostomes to mammals represent a series of increasingly advanced forms, with cyclostomes being relatively close to the ancestral vertebrate source – when their morphology is studied in detail the evidence for the sequence is far less convincing.

For example, it has always been traditionally considered that the morphology of the vertebrate heart and aortic arches in fish, lungfish, amphibia, reptile and mammals form a clear series. However, the sequence is very much a broken one, and it is doubtful to what

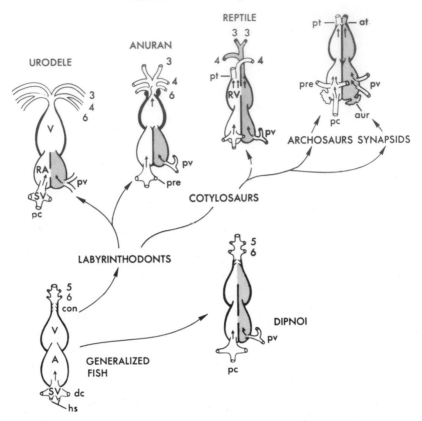

Figure 5.2: The Heart and Aortic Arches in the Major Vertebrate Types. *Note that the morphology of each type is quite distinct. In the mammal the aorta, the vessel which carries the oxygenated blood away from the heart is derived from the fourth left aortic arch while in the reptile by the fourth right aortic arch. Note also that in the case of the lungfish the blood returns to the heart via a pulmonary vein as it does in all other air-breathing vertebrates. The lungfish heart is not intermediate between that of fish and amphibia. Indeed, in its possession of a partially divided ventricle it may be considered closer to the reptilian condition than any amphibian.* (from Kent)[20]

Modifications of the heart that result in increased separation of oxygenated and deoxygenated blood. The parts of the heart shown are **A**, atrium; **RA**, right atrium; **V**, ventricle; **LV**, left ventricle; **RV**, right ventricle; **SV**, sinus venosus; **con**, conus arteriosus; **aur**, auricle of mammalian heart. **3** to **6**, Third to sixth aortic arches. Other vessels are **at**, aortic trunk; **dc**, common cardinal vein; **hs**, hepatic sinus; **pc**, postcava; **pre**, precava (common cardinal vein); **pv**, pulmonary veins; **pt**, pulmonary trunk. Gray chambers contain chiefly, or only, oxygenated blood.

extent it really gives evidence of being a sequence (see Figure 5.2). Take one section of the traditional sequence: amphibia → reptile → mammal. There are many detailed aspects of their comparative anatomy which do not support it, for example, the aortic arches. The major vessel leaving the left ventricle in a reptile, which is the major vessel carrying aereated blood from the heart, is formed from the fourth right aortic arch, while in a mammal it is derived from the left aortic arch (see Figure 5.3). Instead of arranging them in a sequence amphibia → reptile → mammal we might just as easily arrange them circumferentially with reptile and mammal equidistant from amphibia. Moreover, even though the sequence fish → amphibia → reptile is slightly more convincing, there is again much justification for considering the cardiovascular system in amphibia and reptile as unique specializations approximately equidistant from a typical fish (see Figure 5.3).

The only section of the series that is convincing in any sense is the sequence fish → lungfish → amphibia. But as we have already seen, although the lungfish does seem to be intermediate in an overall sense between fish and amphibia, its organ systems are not strictly transitional. Its aortic arches are essentially like those of any fish while the mode of return of aereated blood from the lungs is essentially amphibian, and the heart is of a highly specialized design which differs significantly from any fish or amphibian and certainly cannot be construed to any degree as ancestral to any modern amphibian.

But if there is some hint of a sequence in the case of the aortic arches, it is hopeless trying to arrange vertebrate egg cells, and the pattern of cell division in the earliest stages of embryology up to the formation of the blastula and beyond, into any sort of convincing sequence (see Figure 5.4). In some ways, mammalian eggs are closer in their initial pattern of development to those of a frog than to any reptile.

A final point which should be borne in mind when judging whether the vertebrate series represents a genuine evolutionary sequence is that the environments for which the successive types are adapted are clearly sequential in character: aquatic → semi-aquatic → terrestrial. It would be very surprising if systems, such as the respiratory and cardiovascular which must of necessity satisfy severe environmental constraints, did not reflect the environmental sequence in terms of their general adaptive design. The fact that other aspects of their biology, such as the events of early embryology

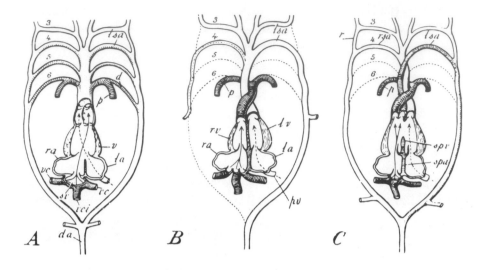

Figure 5.3: The Arrangement of the Heart and Aortic Arches in the Amphibia, Reptiles and Mammals. *Note that the arrangement of the heart and aortic arches of the reptile and mammal can be interpreted in typological terms as being equidistant from the condition in amphibia.* (from Romer)[21]

Diagram of the heart and aortic arches in tetrapods. *A*, Amphibian; *B*, a mammal; *C*, typical modern reptiles; Ventral views; the heart (sectioned) is represented as if the chambers were arranged in the same plane; the dorsal ends of the arches are arbitrarily placed at either side. Solid arrows represent the main stream of venous blood; arrows with broken line the blood coming from the lung. Vessels apparently carrying aerated blood are unshaded; those which appear to contain venous blood, hatched.

 The two vessels at the top of each figure are the internal carotid (laterally) and external carotid (medially). In amphibians without a ventricular septum the two blood streams are somewhat mixed; subdivision of the arterial cone tends to bring about partial separation, but some venous blood is returned to the dorsal aorta. In mammals ventricular separation is complete, the arterial cone subdivided into two vessels, and the arches are reduced to the left systemic and pulmonary.

d, ductus Botalli; *da*, dorsal aorta; *la*, left atrium; *lsa*, left systemic arch; *lv*, left ventricle; *p*, pulmonary artery; *pv*, pulmonary vein; *r*, portion of lateral aorta remaining open in some reptiles; *ra*, right atrium; *rsa*, right systemic arch; *rv*, right ventricle; *spa*, interatrial septum; *spv*, interventricular septum; *sv*, sinus venosus; *v*, ventricle; *vc*, anterior vena cava; *vci*, posterior vena cava.

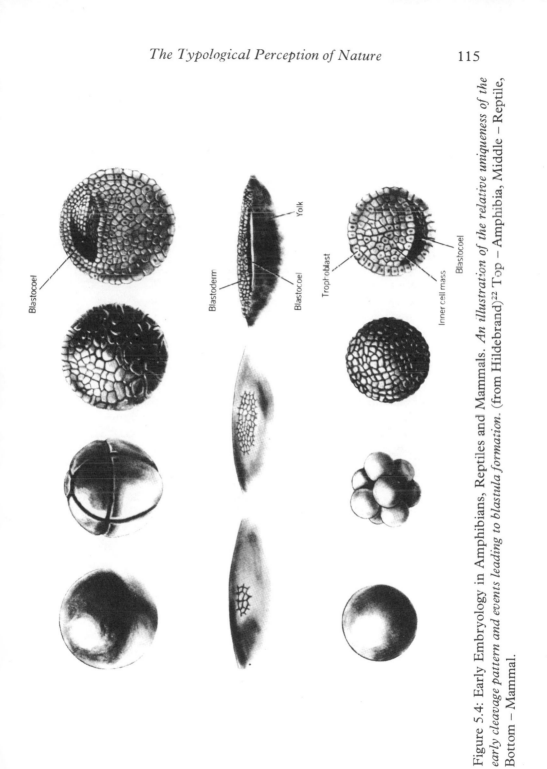

Figure 5.4: Early Embryology in Amphibians, Reptiles and Mammals. *An illustration of the relative uniqueness of the early cleavage pattern and events leading to blastula formation.* (from Hildebrand)[22] Top – Amphibia, Middle – Reptile, Bottom – Mammal.

which are not under such direct environmental constraints exhibit no obvious sequential pattern, suggests that where we do see a sequential arrangement this could just as well be taken to be the result of adaptive necessity rather than as evidence of natural evolutionary relationships.

Another case of a supposed sequential arrangement is the primate series from tree shrew through lemur to monkey and finally to the apes and man. Thomas Huxley referred to it in his *Man's Place in Nature*:[23]

> Perhaps no order of mammals presents us with so extraordinary a series of gradation as this – leading us insensibly from the crown and summit of the animal creation down to creatures, from which there is but a step, as it seems, to the lowest, smallest, and least intelligent of the placental mammals.

But the "sequence" becomes far less convincing when it is critically examined. As primatologist John Beuttner Janusch cautions:[24]

> . . . it is not at all clear whether any living primate populations can be taken to represent truly transitional forms or new forms which led to the next most advanced stages. An example will make this clear. There is no known living member of the order which represents the first erect, bipedal hominid. Nor does there exist any form which is transitional between the quadrupedal primates and the bipedal, ie, a form which exhibits the bony and neuromuscular changes in the pelvic girdle which led to the first erect, bipedal type.

Similarly, in the case of most of the other classic sequential arrangements such as, for example, the land plant series from the "primitive" mosses (the Bryophyta) through the ferns (the Pteridophyta) to the advanced conifers and flowering plants (the Spermatophyta), it invariably turns out that on critical examination the evidence for sequence is vague and ill defined. Moreover, in every case the divisions between each of the successive classes such as those between reptiles and mammals, between lemurs and monkeys, between mosses and ferns, are absolutely clear cut. Whatever conclusion we wish to draw from such "sequential" arrangements they provide no evidence whatsoever for believing that one type of organism was ever converted gradually into another through a series of intermediate forms as Darwin's model of evolution requires.

All in all, the empirical pattern of existing nature conforms remarkably well to the typological model. The basic typological axioms – that classes are absolutely distinct, that classes possess unique diagnostic characters that these diagnostic characteristics are present in fundamentally invariant form in all the members of a class – apply almost universally throughout the entire realm of life. Consequently, the isolation of classes is invariably absolute and transitions to particular character traits are invariably abrupt and the phenomenon of discontinuity ubiquitous throughout the living kingdom.

Even if a number of species were known to biology which were indeed perfectly intermediate, possessing organ systems that were unarguably transitional in the sense required by evolution, this would certainly not be sufficient to validate the evolutionary model of nature. To refute typology and securely validate evolutionary claims would necessitate hundreds or even thousands of different species, all unambiguously intermediate in terms of their overall biology and in the physiology and anatomy of all their organ systems.

The philosophical basis of typology may have been metaphysical nonsense but the fact is that much of the existing pattern of diversity in nature fits very exactly the typological model, and it is easy to see how Cuvier and Agassiz could have seen in the pattern of nature what they took to be irrefutable evidence in favour of their anti-evolutionary stand. The rejection of evolution by the leading biologists of the nineteenth century was not a retreat from empiricism; they simply saw no evidence for a sequential order to the pattern of nature – and this they deemed to be essential if anyone was to induce the concept of organic evolution from the facts of biology. If anyone was chasing a phantom or retreating from empiricism it was surely Darwin, who himself freely admitted that he had absolutely no hard empirical evidence that any of the major evolutionary transformations he proposed had ever actually occurred. It was Darwin, the evolutionist, who admitted in his letter to Asa Gray, that one's "imagination must fill up the very wide blanks."[25]

NOTES

1. Agassiz, L. (1863) *Methods of Study in Natural History,* Ticknor and Fields, Boston, see preface.
2. Romer, A.S. (1977) *The Vertebrate Body,* W.B. Saunders Co. Philadelphia, from Figure 55, p81.

3. Mayr, E. (1970) *Populations, Species and Evolution*, Harvard University Press, Cambridge, Mass, p 4.
4. Darwin, F. (1918) *The Life and Letters of Joseph Dalton Hooker*, 2 vols, John Murray, London, vol 2, pp 302–03.
5. Coleman, W. (1964) *Georges Cuvier: Zoologist*, Harvard University Press, Cambridge, Mass, p 171.
6. Rudwick, M. J. S. (1972) *The Meaning of Fossils*, 2nd ed, Neale Watson Academic Publications, New York, p 241.
7. Hull, D. L. (1973) *Darwin and his Critics*, Harvard University Press, Cambridge, Mass, p 450.
8. Cuvier, G. (1829) *Revolutions of the Surface of the Globe*, English edition, Whittaker, Treacher and Arnot, London, p 63.
9. ibid, p 72.
10. ibid, p 60.
11. Rudwick, op cit, p 227.
12. Owen, R. (1860) "Darwin on the Origin of Species", *Edinburgh Review*, April 1860, 11: 487-532; reprinted in Hull, op cit, p 211.
13. Bronn, H. G. (1860) "Review of the Origin of Species", *Neues Jahrbuch fur Mineralogie*, 112-116; translation in Hull, op cit, p 124.
14. Pictet, J. F. (1860) *Archives des Sciences de la Bibliotheque Universelle*, 3: 231–55; translation in Hull, op cit, p 146.
15. Hull, op cit, p 149.
16. ibid, p 146.
17. Gould, S. J. (1977) "Eternal Metaphors of Palaeontology" in *Patterns of Evolution*, ed. A. Hallam, Elsevier Scientific Pub Co, Amsterdam, pp 1–26, see p 7.
18. Satir, P. (1974) "How Cilia Move", *Scientific American*, 231 (4): 44–52, see p 45.
19. Frey-Wyssling, A. (1973) *Comparative Organellography of the Cytoplasm*, Springer-Verlag, New York, pp 65–66.
20. Kent, G.C. (1973) *Comparative Anatomy of the Vertebrates*, C.V. Mosby and Co, St Louis, from Figure 13.6, p 235.
21. Romer, op cit, from Figure 332, p 422.
22. Hildebrand, M. (1974) *Analysis of Vertebrate Structure*, John Wiley and Sons, New York, from Figure 5.1, p 86.
23. Huxley, T. H. (1876) "Man's Place in Nature" in Huxley, T. H. (1894) *Collected Essays of T. H. Huxley*, 9 vols, Macmillan and Co, London, vol 7, p 146.
24. Buettner-Janusch, J. (1963) "An Introduction to the Primates" in *Evolutionary and Genetic Biology of Primates*, 2 vols, ed J. Buettner-Janusch, Academic Press, New York, vol 1, p 3.
25. Darwin, C. (1858) in a letter to Asa Gray, 5 September 1857, *Zoologist*, 16: 6297–99, see p 6299.

CHAPTER 6

The Systema Naturae from Aristotle to Cladistics

That great and universal feature in the affinities
of all organic beings namely their subordination
in group under group.

Whenever classification schemes are drawn up for phenomena which
fall into a continuous or obviously sequential pattern – such as
climatic zones from the arctic to the tropics, subspecies in a circumpolar overlap, the properties of atoms in the periodic table, series of
fossil horses, or wind strengths from a breeze to a hurricane – class
boundaries are bound to be relatively arbitrary and indistinct. Most
of the classes defined in such schemes are inevitably partially inclusive
of other classes, or, in other words, fundamentally intermediate in
character with respect to adjacent classes in the scheme. Consequently,
when such schemes are depicted in terms of Venn diagrams, most of
the classes overlap and the schemes overall have a disorderly appearance.

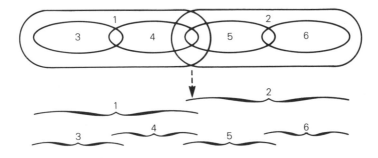

A quite different type of classification system is termed hierarchic,
in which there are no overlapping or partially inclusive classes, but
only classes inclusive or exclusive of other classes. Such schemes

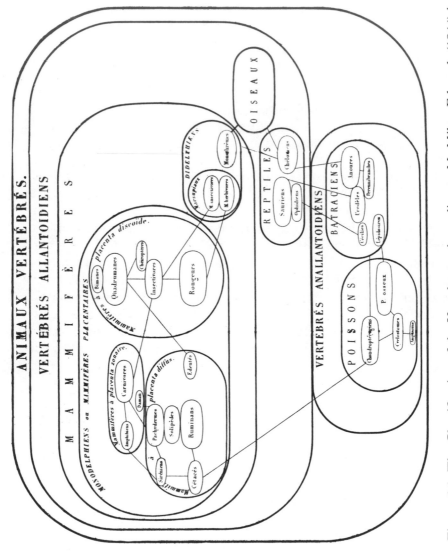

Figure 6.1: The Classification of the Vertebrates According to M. Milne-Edwards (1844).[1]

exhibit, therefore, an orderly "groups within groups" arrangement in which class boundaries are distinct and the divisions in the system increase in a systematic manner as the hierarchy is ascended. The absence of any overlapping classes implies the absence of any sort of natural sequential relationships among the objects grouped by the scheme.

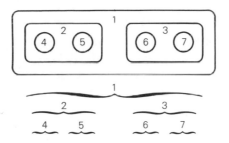

As a rule, overlapping sequential classification schemes occur in the classification of natural phenomena, while hierarchic classification is usually employed in artificial situations, in for example, the subdivisions of a filing system and other similar man-made organizations. The diagram below depicts the subdivisions of modern transport in a hierarchic scheme:

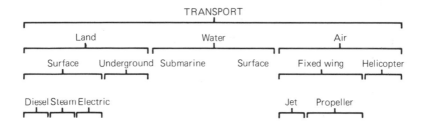

Note how, as in any hierarchic system, the divisions are clear and increase in a systematic way, becoming increasingly intense as the hierarchy is ascended.

It is possible to represent the logic of an ordered hierarchic classification scheme in the form of a branching diagram, or tree, where each node defines the fundamental characteristics of the category grouped by that node.

Such a tree does not imply any sort of natural sequence to the pattern of relationships. The only sequence implied is a theoretical or abstract logical programme whereby a very general concept is successively subdivided into more specific subcategories. The nodes and branches of the tree signify concepts in the mind of the logician and not material entities in the real world. The tree has an ordered appearance with the most specific subcategories, the actual objects grouped by the scheme, occupying peripheral positions as its circumference. Every ordered hierarchial classification system may be reduced to such a logic tree and in every case all the particular objects grouped by the scheme will always be circumferentially arranged at the very periphery of the tree. Groups of objects identified in the scheme are related as sisters or cousins, but never in sequential terms. The tree makes explicit the fact that, where a pattern of relationships is reducible to a highly ordered hierarchic system, the underlying order is fundamentally discontinuous and non-sequential.

The contrast between hierarchic and overlapping classification schemes could not be more complete. While the divisions in an overlapping scheme are blurred and indistinct, those in a orderly hierarchic scheme are distinct and perfectly systematic, increasing in intensity as the hierarchy is ascended. Whereas overlapping schemes imply natural sequential relationships, a hierarchic scheme implies artificial logical relationships of a non-sequential sisterly kind.

Biological classification is basically the identification of groups of organisms which share certain characteristics in common and its beginnings are therefore as old as man himself. It was Aristotle who first formulated the general logical principles of classification and founded the subject as a science. His method employed many of the principles which are still used by biologists today. He was, for example, well aware of the importance of using more than one characteristic as a basis for identifying classes, and he was also aware of the difficult problem which has bedevilled taxonomy ever since: that of selecting the characteristics to be used and weighing their relative significance. His knowledge of comparative anatomy was remarkable for his time. He was aware, for example, of the difference between homologous resemblance, which implies close relationship, and analogous resemblance, which does not.

It is in Aristotle's classification scheme that can first be seen the beginning of the notion of nature as an ordered hierarchial system in which each group is subordinated within a less inclusive group until the highest categories of the hierarchy are attained. Altogether, Aristotle included five hundred different species in his classification.[2] He recognized a primary cleavage between the animals with red blood (roughly our vertebrates) and those without red blood (roughly our invertebrates). The four major subgroups of those possessing blood were: the live-bearing or viviparous quadrupeds which included man and whales (corresponding roughly to our class mammalia); the birds; the egg laying or oviparous quadrupeds in which he included the snakes (class reptilia today); the fishes which he sometimes split into two groups: the bony fish and the cartilaginous fish; the sharks and rays.

Aristotle also divided the bloodless (invertebrates) groups into four subgroups. These were: the cephalopods which included the octopus and squid; the higher Crustacea, the crabs and the lobsters; the insects; the shellfish in which he included a diverse group, molluscs with shells, barnacles and sea urchins.

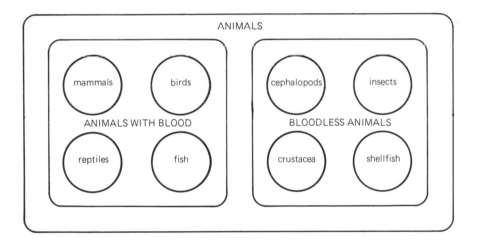

The perception that the pattern of nature conformed to an ordered hierarchic system reappeared with the birth of modern biology in the eighteenth-century and was expressed clearly by Linnaeus in his famous *Systema Naturae*. As knowledge of biology increased during the late eighteenth century and early nineteenth century, the under-lying hierarchic order of nature was increasingly reaffirmed by nearly all the great naturalists and biologists of the time (with one or two exceptions such as Lamarck). As Lovejoy remarks in his *Great Chain of Being*.[3]

> Thus it was that from the end of the sixteenth to the end of the eighteenth century, the project of distributing all living beings, animal or vegetable, into a hierarchy of collective units enclosed one within another, gained such a hold upon naturalists, that it finally seemed to them the formulation of their scientific task.

By the mid-nineteenth century, when knowledge of comparative anatomy was virtually complete, the idea that the pattern of life was reducible to highly ordered groups within groups was almost universally acknowledged. An example of a perfectly ordered hierarchic scheme which typified the perception of nature of most biologists just before the advent of evolution is that of M. Milne Edwards shown in Figure 6.1.

Note how every class is perfectly distinct and totally inclusive or exclusive of other classes. There is a complete absence of any partially inclusive or intermediate classes indicative of sequential relationships. The scheme expresses succinctly the pre-evolutionary belief that nature's order was fundamentally non-sequential.

Even with the rise of evolution and the rejection of the whole metaphysical basis of typology, the perception of nature's order as fundamentally hierarchic persisted largely unchanged. Taxonomists remained just as committed to the highly ordered 'groups within groups' classification schemes as their typological predecessors. As Mayr remarks:[4]

> One would expect a priori that such a complete change of the philosophical basis of classification would result in a radical change of classification, but this was by no means the case. There was hardly any change even in method before and after Darwin, except that the "archetype" was replaced by the common ancestor.

And along similar lines Simpson comments:[5]

> From their classifications alone, it is practically impossible to tell whether zoologists of the middle decades of the nineteenth century were evolutionists or not. The common ancestor was at first, and in most cases, just as hypothetical as the archetype, and the methods of inference were much the same for both, so that classification continued

to develop with no immediate evidence of the revolution in principles. . . . the hierarchy *looked* the same as before even if it meant something totally different.[6]

Haeckel's famous evolutionary tree (see Figure 2.1) illustrates that all the major classes identified by the typologists and even their sisterly relationships survived the advent of evolutionary biology. It is obvious from the way Haeckel drew up his tree that he considered most groups to be related as cousins, not as ancestors and descendants. Some groups are near ancestors, but only *near*; no groups are directly ancestral to other groups. The overall form of Haeckel's tree with the long branches representing hypothetical ancestral pathways is a remarkable concession to the reality of discontinuity and the absence of sequence in nature. Despite the infusion of the spirit of continuity the perception of a hierarchic order persisted.

Even today zoologists find it impossible to relate the major groups of organisms in any sort of lineal or sequential arrangement. This can be seen in the evolutionary trees of the animal kingdom (see Figure 6.2) which were drawn up recently by contemporary zoological authorities. Not only are most groups placed peripherally, giving the trees a circumferential appearance, but many groups are so isolated and unique and of such doubtful affinities that there is complete disagreement as to where they should be placed in the tree. Notice particularly the very different positions of the groups Mollusca and Plathyhelminthes in each of the four schemes.

Over the past twenty years or so taxonomy has experienced something of a revolution and a variety of new methodologies, some using computer technology have been developed and applied to the age old problem of determining the true pattern of nature. Surprisingly, considering the omnipresence of evolutionary theory in so many areas of biology, many of these new taxonomic techniques aim to determine the order of nature in a way which is as free as possible from any *a priori* evolutionary bias.

One of these new schools is often referred to as phenetic. Its aims were described by E. Mayr in a recent article in *Science*. As he points out, the advocates of this school proposed:[7]

. . . in order to make the method more objective, that every character be given equal weight, even though this would require the use of large numbers of characters (preferably well over a hundred). In order to

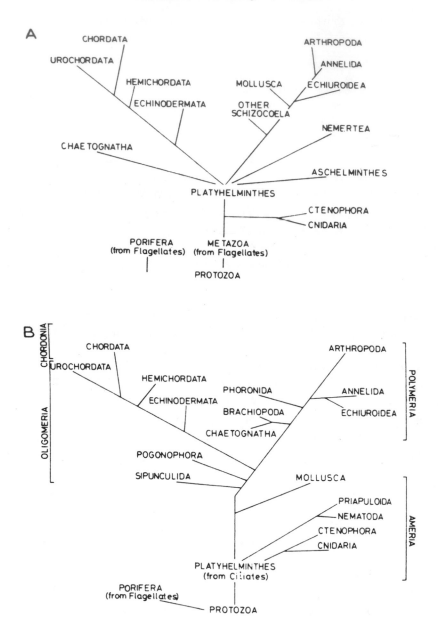

Figure 6.2: Four Evolutionary Trees. *Four models of the phylogeny of meta-zoan phyla according to four leading modern authorities: A – Hyman; B – Hadzi; C – Salvini Plawen; D – Jagersten.* (from Valentine)[8]

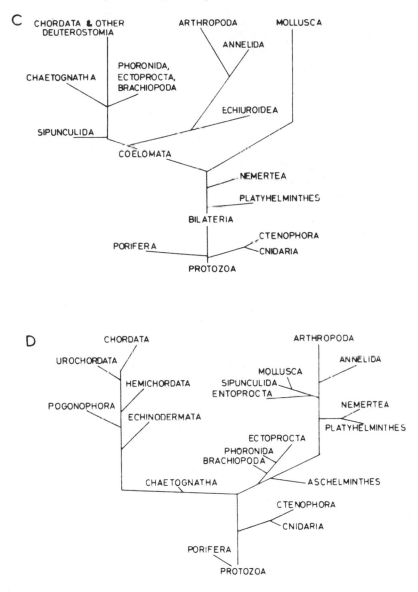

reduce the values of so many characters to a single measure of "overall similarity," each character is to be recorded in numerical form. Finally, the clustering of species and their taxonomic distance from each other is to be calculated by the use of algorithms that operationally manipulate characters in certain ways, usually with the help of computers. The resulting diagram of relationship is called a phenogram. The calculated phenetic distances can be converted directly into a classification.

The diagram below shows a very simple phenogram and its derived classification scheme depicting the relationship of five hypothetical species A, B, C, D and E.

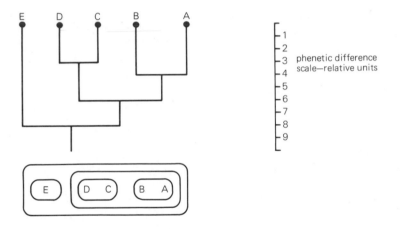

Note species C and D differ by three phenetic units, A and B by four units. The groups A and B and C and D differ by six units while E differs from the other groups by eight units.

The phenetic methodology may be more objective than some of the older approaches to taxonomy, but the classification patterns which result from its application again exhibit a typical orderly hierarchic form.

Perhaps the most influential of these new schools of taxonomy is now widely known as cladistics. Cladistic analyses tend to generate classification schemes from which most of the intermediates or ancestral groups traditionally cited by evolutionary biology as "evidence" are absent or revealed to have no objective basis. As a rule, cladistic procedures tend to depict nature in strikingly non-sequential terms.

Although a detailed description of cladistic methodology is beyond the scope of this chapter the following very simplified account conveys something of the essence of its approach. Cladism takes no account at all of any evolutionary claim regarding the geneology or derivation of any particular species or group. Cladists aim only to discover the pattern of nature as it actually is. In generating a classification scheme the cladist sets out to construct a branching diagram or cladogram which depicts, in the most economical or parsimonious way, the distribution pattern of unique shared characteristics (homologies) among a group of organisms. A simple cladogram is shown below. Each node in the cladogram symbolizes the unique homologies shared by the organisms grouped by that node:

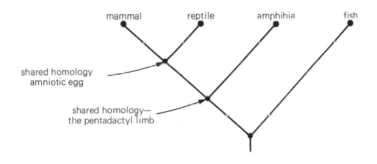

Suppose we wish to generate a cladogram depicting the relationships of the following six organisms, a man, a kangaroo, a hawk, a lizard, a frog and a salamander. We might start the process with any species, say in this case, man. We begin by establishing the nearest relative or sister species of man. This sister species will be one that shares with him some features or homologies which are not shared by any of the other species. In our example, the nearest neighbour to man is obviously the kangaroo which possesses, like man, a whole suit of shared features or homologies including hair and mammary glands. They share many other features, such as the pentadactyl limb and the amniotic egg, but these are, in cladistic jargon, primitive shared features, in other words, homologies shared by many of the other more distantly related species under consideration and hence not relevent to our search for the nearest neighbour of man. We can now relate man, kangaroo and the remaining species thus:

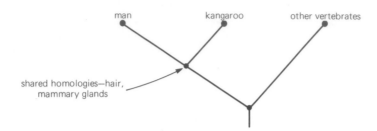

Continuing the process after a series of successive outgroup comparisons, we finally arrive at a cladogram for all the six species:

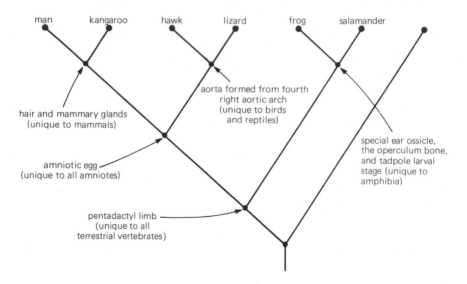

From this cladogram we can derive the following hierarchic classification scheme:

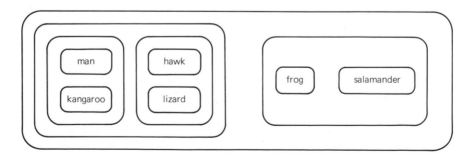

Like many other approaches to classification, yet again, cladistic procedures invariably result in strikingly hierarchic patterns.

While, no doubt, there will always be controversy over various issues in taxonomy, it is significant that since the time of Aristotle the living world has been generally perceived by taxonomists as conforming to a highly ordered hierarchial system of clearly differentiated classes. Moreover, one can trace a consistent theme from the very beginnings of taxonomy that, as biological knowledge has grown, the tendency to perceive nature's order in terms of orderly hierarchic systems of clearly defined classes has grown increasingly intense. Willi Hennig writes:[9]

> In biological systematics repeated attempts have been made to introduce systems other than the hierarchic form. Plate (1914) mentions as an example the ornithological "quinary system" of Kaup (1849), "which is based on the contention that all differences and groupings of vertebrates appear according to the number five." Similar quinary systems have been proposed by other zoologists (Oken, MacLeay, Vigors). Reichenbach proposed a quarternary system. Further examples are given by Stresemann (1951). According to Paramanow (1937) Lubistshev has occupied himself with the possibilities of other biological systems that deviate from the hierarchic system. "Lubistshev calls one such system 'combinative.' It has the appearance of a crystallographic latticework of many dimensions corresponding to the number of independently varying characters." A "periodic system of butterflies" has even been attempted (Barchmeiev 1903–04)! The fact that none of the various nonhierarchic systems has survived suggests that the hierarchic system best expresses the structure of the complex of relations that interconnects all organisms.

While hierarchic schemes correspond beautifully with the typological model of nature, the relationship between evolution and hierarchical systems is curiously ambiguous. Ever since 1859 it has been traditional for evolutionary biologists to claim that the hierarchic pattern of nature provides support for the idea of organic evolution. Yet, direct evidence for evolution only resides in the existence of unambiguous sequential arrangements, and these are never present in ordered hierarchic schemes.

Of course evolutionary biologists do not look for direct evidence in the hierarchy itself but rather argue, as Darwin did, that the hierarchic pattern is readily explained in terms of an evolutionary tree.

Admittedly, the hierarchy is very suggestive of some sort of evolutionary tree. Even pre-Darwinian typologists like Agassiz viewed the hierarchy as evidence for an evolutionary tree, but only in the sense of an abstract branching cladogram. The sort of evolution they conceived was the creative derivation of all the members of a class from the hypothetical archetype which existed in the mind of God. When typologists drew up branching tree diagrams to illustrate the relationships between different species, this did not imply that the members of a class had been derived by natural descent from a common ancestor. None of the nodes or branches of such trees had any real empirical existence; they were 'links' but only in an abstract and ideal sense. As Agassiz in his essay on classification maintained:[10]

> *What we call branches expresses, in fact, a purely ideal connection between animals, the intellectual conception which unites them in creative thought.* It seems to me that the more we examine the true significance of this kind of group, the more we shall be convinced that they are not founded upon material relations.
>
> [*emphasis added*]

The trees of the typologists (see Figure 6.3) merely represented the abstract logic which underlies all hierarchic systems of relationships, the branches and interconnecting nodes being purely theoretical. The fact that all the individual species must be stationed at the extreme periphery of such logic trees merely emphasized the fact that the order of nature betrays no hint of natural evolutionary sequential arrangements, revealing species to be related as sisters or cousins but *never* as ancestors and descendents as is required by evolution. The form of the tree makes explicit the pre-evolutionary view that it is discontinuity and the absence of sequence which is the most characteristic feature of the order of nature.

To the pre-evolutionary biologists the hierarchic order of nature and the astonishing distinctness of each class was a source of considerable wonderment. The question was raised: why should classes be so distinct, why should nature have such an orderly appearance? As Mayr remarks:[11]

> Even in a strictly morphological classification, the assignment of a species to a definite category characterizes it usually as possessing a very definite combination of structures and biological attributes. So perfect indeed was the agreement of taxonomic position and structural

characteristics that it became a source of considerable amazement and speculation among the naturalists in the post-Linnaean period. . . . To explain the orderliness of the natural system, some of the natural philosophers in the first half of the ninetenth century attempted to construct systems on the basis of logical categories, similar to the periodical table of the chemical elements.

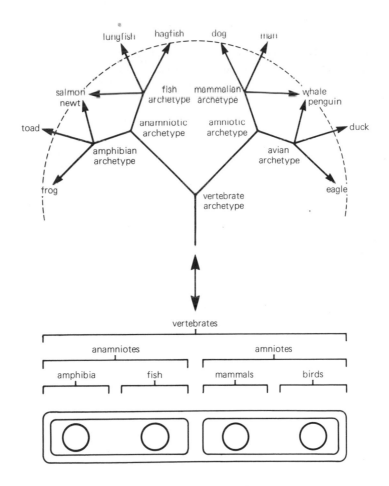

Figure 6.3: The Typological Perception of Nature. *The tree illustrates a typological interpretation of the relationships between four vertebrate classes. Its form reveals why typology leads inevitably to hierarchic classification schemes and, vice versa, how readily hierarchic schemes may be interpreted in typological terms. The tree depicts graphically the fact that for the typologists nature's order was decidedly non-sequential.*

We have interesting evidence that Darwin, especially in his younger years when the idea of evolution was maturing in his mind, also found the orderliness of classification schemes very puzzling. In his book *Darwin on Man*, Gruber points out how much the orderliness troubled Darwin:[12]

> He needed to satisfy himself that the appearance of a jewel-like perfection in the natural order . . . concealed a rampant irregularity beneath the surface, and that this was true at every level of classification, species, genera, orders, families, and classes. Thus he wrote, "Organized beings represent a tree, irregularly branched; some branches far more branched, – hence genera." . . . And he had to show that the natural order did not exhibit a mystical and miraculous regularity and perfection but was instead an irregularly branching system.

While the hierarchic order may not be jewel-like in its perfection, it is not easy to see how a random evolutionary process could have generated such a highly ordered pattern. Consider the hierarchic scheme below which depicts the relationship between four organisms, A, B, C and D, possessing unique character traits α β γ δ ε π φ:

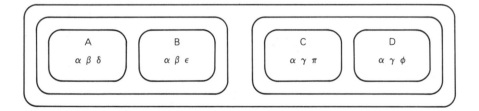

Such a scheme can be readily derived from an evolutionary branching tree which shows the descent of the four organisms from a common hypothetical ancestor Z, and the time of acquisition of the diagnostic character traits (see diagram opposite).

As can be seen in the diagram opposite, if the pattern is to be ordered, one condition that must be met is that character traits once acquired during the course of evolution can never subsequently be lost or transformed in any radical sense and that the acquisition of new character traits must leave, therefore, previously acquired character traits essentially unchanged – to presume, in other words, that evolution is a conservative process such that each phylogenetic lineage gains a succession of what are essentially immutable character

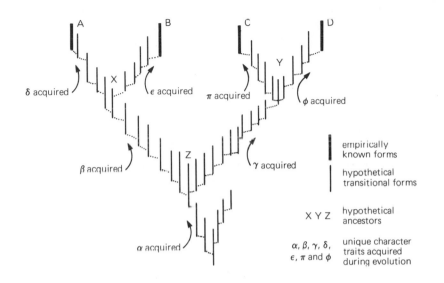

traits. Only if diagnostic character traits remain essentially immutable in all the members of the group they define is it possible to conceive of a hierarchic pattern emerging as the result of an evolutionary process.

For example, in the tree above, if character α was lost during the evolution of $X \rightarrow B$ and $Y \rightarrow C$, the hierarchy would be unrecognizable leaving the character traits thus:

A	B	C	D
α β δ	β ϵ	γ π	α γ ϕ

It is surely a matter of debate as to what extent the existence of invariant character traits is really compatible with the notion of evolution as a random radical process of change. For if it is true, as the Darwinian model of evolution implies, that all the character traits of living things were gained in the first place as a result of a gradual random evolutionary process, then why should they have remained subsequently so fundamentally immune to that same process of change, especially considering that many diagnostic character traits are only of dubious adaptive significance? It was precisely this fundamental constancy of the unique character traits, or homologies, of every defined taxon which led nineteenth-century biology to the theory of types!

Apart from the fact that the idea that certain characteristics of living organisms are invariant is purely *ad hoc* and does not flow naturally from the Darwinian concept of evolution as a random and essentially undirected process of change, it is not at all clear if such 'restrictions' on the course of evolutionary change would be compatible with the radical sorts of evolutionary transformations which must have been involved in converting an amoeba into a mammal or a fish into a bird.

There is another stringent condition which must be satisfied if a hierarchic pattern is to result as the end product of an evolutionary process: *no ancestral or transitional forms can be permitted to survive.* This can be seen by examining the tree diagram above on page 135. If any of the ancestors X, Y and Z, or if any of the hypothetical transitional connecting species stationed on the main branches of the tree, had survived and had therefore to be included in the classification scheme, the distinctness of the divisions would be blurred by intermediate or partially inclusive classes and what remained of the hierarchic pattern would be highly disordered.

In the *Origin*, Darwin saw extinction as playing an important role in isolating and widening the gaps between different types.[13]

> Extinction, as we have seen in the fourth chapter, has played an important part in defining and widening the intervals between the several groups in each class. We may thus account for the distinctness of whole classes from each other – for instance, of birds from all other vertebrate animals – by the belief that many ancient forms of life have been utterly lost, through which the early progenitors of birds were formerly connected with the early progenitors of the other and at that time less differentiated vertebrate classes.

But surely no purely random process of extinction would have eliminated so effectively all ancestral and transitional forms, all evidence of the trunk and branches of the supposed tree, and left all remaining groups: mammals, cats, flowering plants, birds, tortoises, vertebrates, molluscs, hymenoptera, fleas and so on, so isolated and related only in a strictly sisterly sense.

In the final analysis the hierarchic pattern is nothing like the straightforward witness for organic evolution that is commonly assumed. There are facets of the hierarchy which do not flow naturally from any sort of random undirected evolutionary process. If the hierarchy suggests any model of nature it is typology and not evolution. How much easier it would be to argue the case for evolution if all nature's divisions were blurred and indistinct, if the *systema naturae* was largely made up of overlapping classes indicative of sequence and continuity.

The inherent contradiction between an orderly hierarchic pattern and a random evolutionary process, which was apparent to many biologists in the early nineteenth century, persisted after 1859, although awareness of it was largely restricted to the small circle of professional taxonomists where, as Simpson admits, typological concepts continued to influence "all schools of taxonomy including some that usually oppose typology in principle".[14] It has only been over the past two decades, with the adoption of new methodologies which have subsequently revitalized and popularized the science of classification, that the conflict between hierarchy and evolution has re-emerged and come to the attention of significant numbers of biologists. The re-emergence of the conflict is evidenced today not only in the increasing scepticism being expressed by some of the more radical cladists over many aspects of evolution theory, but also in the increasing resemblance that is developing between the modern cladistic framework and the non-evolutionary perception of pre-Darwinian biology.

Another aspect of cladism that is reminiscent of typology is the renewed importance attached to the defining of diagnostic characters of different groups. This inevitably tends to highlight the distinctness of individual classes and sharpen divisions between them, thereby emphasizing the discontinuous appearance of nature. By its very nature a cladogram draws attention to the fact that all the species grouped under one node possess to the same degree those unique defining characters of their class, so that they are all equally representative of their class and equidistant from all other organisms in terms of these fundamental characteristics. Kangaroo, mouse and man, for example, all equally share the basic suit of mammalian characteristics, hair, mammary glands, etc and stand therefore in respect to these features equidistant from all other vertebrates species.

In emphasizing the distinctness of biological classes, in its seeking sisterly relationships and the consequent tendency to eliminate sequential arrangements, in its increasing stress that ancestors cannot be empirically known, cladism is conjuring up a sequence-free perception of nature very similar to the discontinuous typological model of the early nineteenth century. There is no longer any need to read between the lines to see a form of typology re-emerging in the thinking of many leading zoologists and taxonomists today. Gareth Nelson and Norman Platnick make the confession explicit:[15]

> Since the advent of the so-called New Systematics, it has become popular to deprecate as *"essentialistic"* or *"typological"* the notions that species (and hence groups of them) have defining characters, and that it is the business of systematics to find them. . . . *systematists always have been, are, will be, and should be, typologists.*
>
> *[emphasis added]*

And Colin Patterson of the British Museum of Natural History quotes Nelson as admitting in a personal communication:[16]

> In a way, I think we are merely rediscovering *pre-evolutionary systematics*: or if not rediscovering it, fleshing it out.
>
> *[emphasis added]*

Because of its affinity to typology and the sceptical position adopted by many cladists with regard to much of traditional evolutionary biology, cladism is being viewed with increasing suspicion by many members of the biological establishment. It is the threat to evolutionary biology posed by growing application of cladistic principles that led the journal *Nature* to launch a vitriolic attack against the staff of the Natural History Museum in South Kensington when they introduced cladistic principles in the reorganization of the public galleries.[17] The editors of *Nature* can see where cladism is leading. As Patterson put it recently:[18]

> . . . as the theory of cladistics has developed, it has been realized that more and more of the evolutionary framework is inessential, and may be dropped. The chief sympton of this change is the significance attached to nodes in cladograms . . . in all early work in cladistics, the nodes are taken to represent ancestral species. This assumption has been found to be unnecessary, even misleading, and may be dropped. Platnick (1980) refers to the new theory as "transformed cladistics" and *the transformation is away from dependence on evolutionary theory.*
>
> *[emphasis added]*

Ultimately, evidence for evolution exists and only exists in sequential or ancestor-descendant arrangements. Small wonder then that the evolutionary community is viewing cladism with a growing sense of unease, seeing within it the seeds of an intellectual revolution which could eventually seriously threaten the credibility of the whole evolutionary model of nature. When Wilma George recently described cladism as a *"non-evolutionary* classification"[19] [*emphasis added*] she may not have been too far from the mark. Keith Thompson of Yale University used the term antithetical to describe the relationship of cladism to evolutionary biology:[20]

> No one needs reminding that we are well into a revolutionary phase in the study of evolution, systematics, and the interrelationships of organisms. . . . to the thesis of Darwinian evolution . . . has been added a *new cladistic antithesis* which says that the search for ancesors is a *fool's errand,* that all we can do is determine sister group relationships based on the analysis of derived characters. . . . It is a change in approach that is not easy to accept for, in a sense, it runs counter to what we have all been taught.
>
> [*emphasis added*]

Whatever the future of cladism, the fact that a significant number of biologists in the 1980s are insisting, in the words of Beverly Halstead (no friend of cladism himself), that *"no species can be considered ancestral to any other"*[21] [emphasis added] marks without question a watershed in evolutionary thought.

Agassiz would surely have smiled to see his perception of sequence as essentially absent from nature being reaffirmed by modern cladism. How ironic to think that contemplation of the hierarchic pattern of nature, which Darwin saw as one of his most important allies, has led significant numbers of leading contemporary biologists to conclude, like Patterson: "that much of today's explanation of nature, in terms of neo-Darwinism, or the synthetic theory, may be empty rhetoric."[22] And to insist that as far as ancestors are concerned: "they exist not in nature but in the mind of the taxonomist, as abstractions . . . yet they are always discussed as if they have some reality . . ."[23]

In a sense the antagonism between cladism and evolutionary biology is only the latest manifestation of the inherent contradiction between taxonomy with its distinct divisions and ordered hierarchy and the fundamental need of evolutionary biology to demonstrate the existence of sequence in nature. Cladism is merely making explicit a fact enshrined in classification schemes ever since Aristotle but which has lain dormant for most of the past century – that in the final analysis nature's order is not sequential.

NOTES

1. Milne-Edwards, M. (1844) "Considerations sur quelques principes relatifs a la classification naturelle des animaux", *Ann. Sci. Nat.*, series 3, 1: 65–99, Plate.
2. Wright, S. (1968) *Evolution and the Genetics of Populations*, 4 vols, Chicago University Press, Chicago, vol 1, p 1
3. Lovejoy, A.O. (1961) *The Great Chain of Being*, Harvard University Press, Cambridge, Mass, p 228
4. Mayr, E., Lindley, E.G. and Usinger, R,L. (1953) *Methods and Principles of Systematic Zoology*, McGraw-Hill Book Co, New York, p 5.
5. Simpson, G.G. (1945) "The Principles of Classification and a Classification of Mammals", *Bull. Amer. Mus. of Nat. Hist.*, 85: 1–350, see p 4.
6. Simpson, G.G. (1953) *Life of the Past*, Yale University Press, New Haven, p96
7. Mayr, E. (1981) "Biological Classification: Towards a Synthesis of Opposing Methodologies", *Science*, 214: 510–16, see p 510.
8. Valentine, J.W. (1977) "General Patterns of Metazoan Evolution" in *Patterns of Evolution as Illustrated by the Fossil Record*, ed A. Hallam, Elsevier Scientific Pub Co, Amsterdam, pp 27–57, from Figure 1, p 29.
9. Hennig, W. (1966) *Phylogenetic Systematics*, University of Illinois Press, Illinois, p 15.
10. Agassiz, L. (1857) *Essay on Classification*, reprinted 1962, Harvard University Press, Cambridge, Mass, p 147
11. Mayr et al, op cit, p 41.
12. Gruber, H. E. (1981) *Darwin on Man: A Psychological Study of Scientific Creativity*, 2nd ed, University of Chicago Press, Chicago, p 198.
13. Darwin, C. (1872) *The Origin of Species*, 6th ed (1962) Collier Books, New York, p 432
14. Simpson, G.G. (1961) *Principles of Animal Taxonomy*, Columbia University Press, New York, p 49.
15. Nelson, G. and Platnick, N. (1981) *Systematics and Biogeography*, Columbia University Press, New York, p 329.
16. Patterson, C. (1980) "Cladistics", *Biologist*, 27: 234–40, see p 239.
17. Editorial (1981) "Cladistics and Evolution on Display", *Nature*, 292: 395–6.
18. Patterson, op cit, p 239.
19. George, W. (1982) *Darwin*, Fontana, London, see glossary for definition of cladism.

20. Thompson, K. (1981) "A Radical Look at Fish-Tetrapod Relationships", *Paleobiology*, 7: 153–156, see p153.

21. Halstead, B. (1981) "Halstead's Defence Against Irrelevancy", *Nature*, 292: 403–04, see p403.

22. Patterson, op cit, p238.

23. Patterson (1976) "The Contribution of Paleontology to Teleostean Phylogeny" in *Major Patterns in Vertebrate Evolution*, ed M.K. Hecht, Plenum Press, London, pp579–643, see p623.

CHAPTER 7

The Failure of Homology

We have seen that the members of the same class, independently of their habits of life, resemble each other in the general plan of their organisation. . . . Is it not powerfully suggestive of true relationship, of inheritance from a common ancestor.

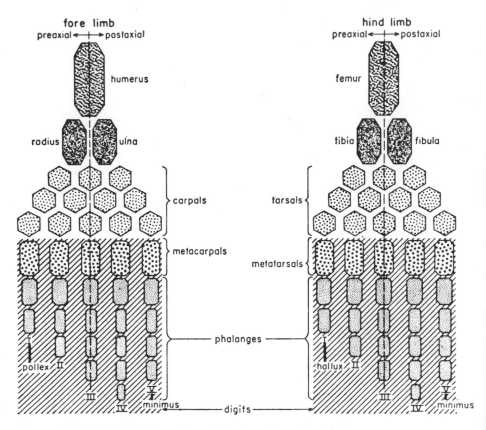

Figure 7.1: The Basic Pentadactyl Design. (from Smith)[1]

Since 1859 the phenomenon of homology has been traditionally cited by evolutionary biologists as providing one of the most powerful lines of evidence for the concept of organic evolution. As in so many other areas of evolutionary thought, no one has ever presented the argument with greater clarity than Darwin himself. It is worth quoting his reasoning at length:[2]

> We have seen that the members of the same class, independently of their habits of life, resemble each other in the general plan of their organisation. This resemblance is often expressed by the term "unity of type"; or by saying that the several parts and organs in the different species of the class are homologous. The whole subject is included under the general term of Morphology. This is one of the most interesting departments of natural history, and may almost be said to be its very soul. What can be more curious than that the hand of a man, formed for grasping, that of a mole for digging, the leg of the horse, the paddle of the porpoise, and the wing of the bat should all be constructed on the same pattern, and should include similar bones, in the same relative positions? How curious it is, to give a subordinate though striking instance, that the hind-feet of the kangaroo, which are so well fitted for bounding over the open plains, – those of the climbing, leaf eating koala, equally well fitted for grasping the branches of trees, – those of the ground-dwelling, insect or root-eating, bandicoots, – and those of some other Australian marsupials, – should all be constructed on the same extraordinary type, namely with the bones of the second and third digits extremely slender and enveloped within the same skin, so that they appear like a single toe furnished with two claws. Notwithstanding this similarity of pattern, it is obvious that the hind feet of these several animals are used for as widely different purposes as it is possible to conceive. The case is rendered all the more striking by the American oppossums, which follow nearly the same habits of life as some of their Australian relatives, having feet constructed on the ordinary plan. Professor Flower, from whom these statements are taken, remarks in conclusion: "We may call this conformity of type, without getting much nearer to an explanation of the phenomenon"; and he then adds "but is it not powerfully suggestive of true relationship, of inheritance from a common ancestor?"

Homology provided Darwin with apparently positive evidence that organisms had undergone descent from a common ancestor. Furthermore, the evolutionary explanation of homology appeared to be one instance where evolution seemed far more plausible than its creationist alternative. On the face of it, it would appear very difficult to explain by a creationist theory the persistence of the so-called pentadactyl pattern in the limbs of all the major terrestrial vertebrates from the first amphibian up to present day forms. Why should a creator be restricted to the same basic pentadactyl design in designing the flipper of a whale or the wing of a flying reptile? Darwin taunted his creationist opponents:[3]

> Nothing can be more hopeless than to attempt to explain this similarity of pattern in members of the same class, by utility or by the doctrine of final causes. The hopelessness of the attempt has been expressly admitted by Owen in his most interesting work on the 'Nature of Limbs'. On the ordinary view of the independent creation of each being, we can only say that so it is: – that it has pleased the Creator to construct all the animals and plants in each great class on a uniform plan: but this is not a scientific explanation.

The phenomenon of homology has remained the mainstay of the argument for evolution right down to the present day. The latest edition of the Encyclopaedia Britannica gives pride of place to homology in discussing the evidence for evolution:[4]

> It must be stressed that Darwin himself never claimed to provide proof of evolution or of the origin of species, what he did claim was that if evolution has occurred, a number of otherwise inexplicable facts are readily explained. The evidence for evolution was therefore indirect. . . . The indirect evident for evolution is based primarily on the significance of similarities found in different organisms. . . . The similarity of plan is easily explicable if all descended with modification from a common ancestor, by evolution, and the term homologous is used to denote corresponding structures formed in this way. . . . In vertebrate animals, the skeleton of the forelimb is a splendid example of homology, in the bones of the upper arm, forearm, wrist, hand, and fingers, all of which can be matched, bone for bone, in rat, dog, horse, bat, mole, porpoise, or man. The example is all the more telling because the bones have become modified in adaptation to different modes of life but have retained the same fundamental plan of structure, inherited from a common ancestor.

The authors of the Penguin *Dictionary of Biology* were certainly understating the case when they described homologous resemblance as "the main concept of evolutionary comparative anatomy",[5] for without underlying homologous resemblance in the fundamental design of dissimilar organisms and organ systems then evolution would have nothing to explain and comparative anatomy nothing to contribute to evolutionary theory.

The validity of the evolutionary interpretation of homology would have been greatly strengthened if embryological and genetic research could have shown that homologous structures were specified by homologous genes and followed homologous patterns of embryological development. Such homology would indeed be strongly suggestive of "true relationship; of inheritance from a common ancestor". But it has become clear that the principle cannot be extended in this way. Homologous structures are often specified by non-homologous genetic systems and the concept of homology can seldom be extended back into embryology. The failure to find a genetic and embryological basis for homology was discussed by Sir Gavin de Beer, British embryologist and past Director of the British Museum of Natural History, in a succinct monograph *Homology, an Unresolved Problem.*[6]

The earliest events leading from the first division of the egg cell to the blastula stage in amphibians, reptiles and mammals are illustrated in Figure 5.4. Even to an untrained zoologist it is obvious that neither the blastula itself, nor the sequence of events which lead to its formation, is identical in any of the three vertebrate classes shown. The differences become even more striking in the next major phase in embryo formation – gastrulation. This process involves a complex sequence of relative cell movements whereby the cells of the blastula rearrange themselves, eventually resulting in the transformation of the blastula into the intricate folded form of the early embryo, or gastrula, which consists of the three basic germ cell layers: the ectoderm, which gives rise to the skin and the nervous system; the mesoderm, which gives rise to muscle and skeletal tissues; and the endoderm, which gives rise to the lining of the alimentary tract as well as to the liver and pancreas. No one doubts that gastrulation and the gastrula are homologous in all vertebrates, yet the way the gastrula is formed and particularly the positions in the blastula of the cells which give rise to the germ layers and their migration patterns during gastrulation differ markedly in the different vertebrate classes. In some ways the egg cell, blastula and gastrula stages in the different vertebrate classes are so dissimilar that, were it not for the close resemblance in the basic body plan of all adult vertebrates, it seems unlikely that they would have been classed as belonging to the same phylum. There is no question that, because of the great dissimilarity of the early stages of embryogenesis in the different vertebrate classes, organs and structures considered homologous in adult vertebrates cannot be traced back to homologous cells or regions in the earliest stages of embryogenesis. In other words, homologous structures are arrived at by different routes.

Even after gastrulation the sites from which homologous structures are derived are different in different vertebrate classes. As De Beer points out, structures as obviously homologous as the vertebrate alimentary canal are formed from quite different embryological sites in different vertebrate classes. The alimentary canal is formed from the *roof of the embryonic gut* cavity in the sharks, from *the floor* in the lamprey, from *roof and floor* in frogs, and from the *lower layer of the embryonic disc*, the blastoderm, in birds and reptiles.[7] Another class of organs considered strictly homologous are the vertebrate forelimbs, yet they generally develop from different body segments in different vertebrate species. The forelimbs develop from the trunk segments 2, 3, 4 and 5 in the newt, segments 6, 7, 8 and 9 in the lizard and from segments 13, 14, 15, 16, 17 and 18 in man.[8] It might be argued that they are not strictly homologous at all! Similarly, the position of the occipital arch relative to body segmentation varies widely in different vertebrate species.

The development of the vertebrate kidney appears to provide another challenge to the assumption that homologous organs are generated from homologous embryonic tissues. In fish and amphibia the kidney is derived directly from an embryonic organ known as the mesonephros, while in reptiles and mammals the mesonephros degenerates towards the end of embryonic life and plays no role in the formation of the adult kidney, which is formed instead from a discrete spherical mass of mesodermal tissue, the metanephros, which develops quite independently from the mesonephros. Even the ureter, the duct which carries the urine from the kidney to the bladder, is formed in a completely different manner in reptiles and mammals from the equivalent duct in amphibia.

A further example is provided by the development of the two unique membranes, the amniotic and allantoic, which surround the growing embryo in reptiles, birds and mammals. These membranes are considered to be strictly homologous in all the vertebrate groups in which they occur, but in mammals the processes which lead to their formation and the cells from which they are derived differ completely from those in reptiles and birds. In De Beer's words:

> It does not seem to matter where in the egg or the embryo the living substance out of which homologous organs are formed comes from. Therefore, *correspondence between homologous structures cannot be pressed back to similarity of position of the cells of the embryo or the parts of the egg out of which these structures are ultimately differentiated.*
>
> [*emphasis added*]

In the same article De Beer goes on to describe in detail an interesting case which illustrates that even actual developmental mechanisms, by which apparently homologous structures are formed during embryogenesis, may not be homologous at all:

> It was a problem to know why the lens of the vertebrate eye, which develops from the epidermis overlying the optic cup, should develop exactly in the "right" place, and fit into the optic cup so perfectly, until it was discovered that the optic cup is itself an organizer which induces the epidermis to differentiate into a tailor-made lens. At least, this is what it does in the common frog, *Rana fusca*, in the embryo of which, if the optic cup is cut out, no lens develops at all. But in the closely related edible frog, *Rana esculents*, the optic cup can be cut out from the embryo, and the lens develops all the same. It cannot be doubted that the lenses of these two species of frog are homologous, yet they differ completely in the mechanism by which determination and differentiation are brought about.
>
> This is no isolated example. In true vertebrates the spinal cord and brain develop as a result of induction by the underlying organizer; but in the "tadpole larva" of the tunicates, which has a "spinal cord" like the vertebrates, it differentiates without any underlying organizer at all. *All this shows that homologous structures can owe their origin and stimulus to differentiate to different organizer-induction processes without forfeiting their homology.*
>
> [*emphasis added*]

Insect metamorphosis provides many other examples of homologous organs and structures being arrived at by radically different embryogenic routes. The first stage of metamorphosis, shortly following the formation of the pupa, involves what amounts to the virtual dissolution of all the organ systems of the larvae into a veritable soup of fragmented cells and tissues. This dissolution phase is quickly followed by an assembly phase during which all the organ systems – muscular,

nervous and alimentary – of the adult insect are built up from special
embryonic cells which occur either in specific places in the pupa,
known as imaginal buds or discs, or scattered widely in the disinte-
grating tissues of the larva.[9] Detailed comparative studies of the
processes of organ formation in different insect species have revealed
that the ways in which the adult organ systems are formed during
metamorphosis are bewilderingly diverse in different species.

Take, for example, the formation of the alimentary tract. The
lining of the midgut is always replaced during metamorphosis but the
new adult midgut is reformed in some insects from primitive embryonic cells scattered throughout the old larval midgut while in other
species by the migration of special embryonic cells from its posterior
end.[10] The process of reconstruction of the fore- and hindguts also
differs radically in different species. As entomologist Chapman
describes it:[11]

> In Coleoptera (the beetles) the reconstruction of the stomatodoeum
> (the foregut) and the proctodaeum (the hindgut) is carried out by the
> renewed activity of the larval cells without any accompanying cell
> destruction but in Lepidoptera (the butterflies) and Diptera (the flies)
> new structures develop from imaginal rings which are proliferating
> centres at the tips of the foregut and hindgut.

Discussing the fate of the malphighian, or excretory, tubules of the
larva, he continues:[12]

> In Coleoptera the tubules are rebuilt from special cells in the larval
> tubules while in Hymenoptera (the ants and bees) the larval tubules
> break down completely and are replaced by new ones developing from
> the tip of the proctodaeum (the hindgut).

No one would doubt that the alimentary tract and the malphighian
tubules are homologous in all insect species but, again as in the
vertebrate cases cited above, we see that "homology" cannot be
traced back to similar embryogenic processes and events.

The same principle also holds in the case of many sorts of homology
in plants. For example, the seeds of the conifers and the flowering
plants (the angiosperms) are considered to be homologous by most
botanists, and indeed the close resemblance in the structure of
the seeds in both groups is used by taxonomists as one of the key
character traits to classify them together in the major group Sper-

matophyta.[13] Each seed consists of an enclosed egg cell or ovule plus a food store (the endosperm) which surrounds the ovule and supplies nourishment to the growing embryo after fertilization. Yet the way in which the ovule and endosperm are formed profoundly differs in the two groups in a number of important respects.[14]

It appears then that Darwin's usage of the term 'homology', which he defines in the *Origin* as that "relationship between parts which results from their development from corresponding embryonic parts",[15] is, as De Beer emphasizes, just what homology is not.

The evolutionary basis of homology is perhaps even more severely damaged by the discovery that apparently homologous structures are specified by quite different genes in different species. The effects of genes on development are often surprisingly diverse. In the house mouse, nearly every coat-colour gene has some effect on body size. Out of seventeen x-ray induced eye colour mutations in the fruit fly *Drosophila melanogaster*, fourteen affected the shape of the sex organs of the female, a characteristic that one would have thought was quite unrelated to eye colour. Almost every gene that has been studied in higher organisms has been found to effect more than one organ system, a multiple effect which is known as pleiotropy. As Mayr argues in *Population, Species and Evolution:*[16]

> It is doubtful whether any genes that are not pleiotropic exist in higher organisms. Since the primary gene action in multicellular organisms is usually several steps removed from the peripheral phenotypic character, it is obvious that non pleiotropic genes must be rare if they exist at all.

Not only are most genes in higher organisms plieotropic in their influence on development but, as is clear from a wide variety of studies of mutational patterns in different species, the plieotropic effects are invariably species specific.

In Figure 7.2 the multiple effects of one particular gene in the domestic chicken are illustrated. As can be seen, a mutation in this gene causes developmental abnormalities in a variety of systems. Here then is a gene that is involved in the development of some structures unique to birds – air sacs and downy feathers – and of other structures such as lungs and kidneys, which occur in many other vertebrate classes. This can only mean that non-homologous genes are involved to some extent in the specification of homologous structures.

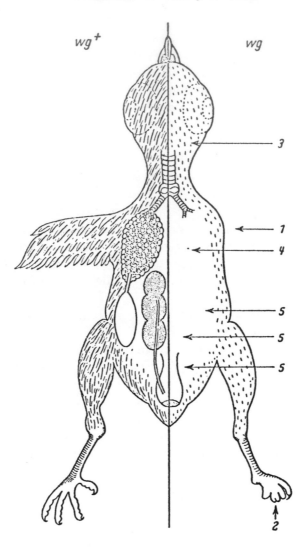

Figure 7.2: A Pleiotropic Gene in the Domestic Fowl. *Left half of the diagram illustrates normal development (wg+) and the right half shows the effect of pleiotropic mutation wingless (wg). The pattern of damage is markedly organ specific. (1) The wings either do not develop at all or they form small stumps; (2) the hind limbs reach full length although the digits often exhibit syndactyly or hyperphalagia; (3) the downy cover remains underdeveloped; (4) the lungs and air sacs are absent although the trachea and extrapulmonary bronchi are normal; (5) the ureter does not grow and fails to induce the development of the kidney.* (from Hadorn)[17]

Another simple example is a gene, again in the domestic fowl, which controls the formation of the crest of feathers and causes a cerebral hernia with upswelling of the skull in the form of a knob to accommodate it. It is difficult to believe that this gene has any homologue in vertebrate species lacking feathers, and yet it is involved in the development of the skull, a feature possessed by all vertebrate species. As De Beer says, "Homologous structures need not be controlled by identical genes and homology of phenotypes does not imply similarity of genotype."[18]

A convincing explanation for the mystifying 'unity of type', the phenomenon of homology that Darwin thought he had so adequately explained by descent from a common ancestor, is probably still a very long way away. With the demise of any sort of straightforward explanation for homology one of the major pillars of evolution theory has become so weakened that its value as evidence for evolution is greatly diminished. The breakdown of the evolutionary interpretation for homology cannot be dismissed as a triviality and casually put aside as a curiosity for, as Sir Alister Hardy reminds us in his book *The Living Stream*:[19]

> The concept of homology is absolutely fundamental to what we are talking about when we speak of evolution – yet in truth we cannot explain it at all in terms of present day biological theory.

The evolutionary interpretation of homology is clouded even further by the uncomfortable fact that there are many cases of 'homologous like' resemblance which cannot by any stretch of the imagination be explained by descent from a common ancestor. The similar pentadactyl design of vertebrate fore- and hindlimbs provides the classic example. We have seen that the *forelimbs* of all terrestrial vertebrates are constructed according to the same pentadactyl design, and this is attributed by evolutionary biologists as showing that all have been derived from a common ancestral source. But the *hindlimbs* of all vertebrates also conform to the pentadactyl pattern and are strikingly similar to the forelimbs in bone structure and in their detailed embryological development. Yet no evolutionist claims that the hindlimb evolved from the forelimb, or that hindlimbs and forelimbs evolved from a common source.

The striking similarity in the design of the fore- and hindlimbs of terrestrial vertebrates is seen in Figure 7.1. The detailed correspondence is remarkable. The proximal part of both the fore- and hindlimb is composed of one main bone, humerus in the arm, femur in the leg. The next section of the limbs is composed of two bones, radius and ulna in the arm, tibia and fibula in the leg. The hand and foot are also based on the same design, with five digits in both hand and foot. The first digit in hand and foot, the thumb and big toe, are both made up of only two small bones; the other digits are made up of three or more. There is no doubt that in terms of evolution the fore- and hindlimbs must have arisen independently, the former supposedly evolving from the pectoral fins of a fish, the latter from the pelvic fins. Here is a case of profound resemblance which cannot be explained in terms of a theory of descent.

The occurrence of the same pentadactyl pattern in the fore- and hindlimbs presents an additional and unrelated challenge to evolutionary biology – that of explaining the independent origin of structures which are incredibly similar in terms of a random accumulation of tiny advantageous mutations. The adult form of the fore- and hindlimbs is not identical in any known vertebrate species. In every case the pentadactyl plan is considerably modified during development, so that the final adaptive form of both limbs is quite different and departs markedly from the basic pentadactyl plan, so much so that in many cases the original pentadactyl design is virtually impossible to detect in the final form of the limb. It seems very unlikely that there could be any adaptive necessity that dictates that there be five digits in both hand and foot or that thumb and big toe be both made up of two phalanges, that the forearm and lower leg be both made of two long bones or that there be only one bone in the upper arm and leg.

We seem forced to propose that during the course of evolution the gradual accumulation of tiny independent and random changes in two independent structures – the pectoral and pelvic fins of a fish – hit on an identical yet apparently arbitrary ground plan for the design of the fore- and hindlimbs of a tetrapod. The problem is even more perplexing considering that neither the initial structures – the pelvic and pectoral fins of a fish – nor the end products of the process – the fore- and hindlimbs of a tetrapod – are in any strict sense identical.

How this complex and seemingly arbitrary pattern was arrived at twice independently in the course of evolution is mystifying. The question is bound to arise: perhaps the analogy between fore- and hindlimbs, the exact adherence to the same pentadactyl plan, is satisfying some, as yet unknown, necessities in the construction of tetrapod limbs? Perhaps only the pentadactyl pattern is, in Cuvier's terms, "compatible" with the vertebrate type. In the context of this unsolved question, it is obviously premature to interpret the occurrence of the pentadactyl pattern in vertebrate forelimbs to descent from a common ancestor. It may in the end reflect a necessity deeply embedded in the developmental logic of vertebrates. Whatever the ultimate explanation for this remarkable pattern turns out to be, there seems little intellectual satisfaction in attributing one case of correspondence to evolution while refusing it in the other. There are many other examples of this sort of the phenomenon, adaptations of great complexity which exhibit very close resemblance in their design but which must have arisen entirely independently. Such examples of convergence led Carter to comment:[20]

> There are many problems in evolution for which our present explanations are inadequate or incomplete. This is certainly one place in which this is so. It is clear that much more work must be done before we have a complete understanding of the process of evolution.

Is it possible that many cases of resemblance in nature which are today classed as homologous, and taken by evolutionary biologists as implying descent from a common origin, may turn out to be merely analogous? There is certainly a long term historical trend which tends to bear this possibility out. Early in his career Linnaeus, for example, mistakenly classed the Cetaceans (the whales) as fish, not realizing that their fish-like shape was only an example of analogous resemblance. Over and over again, as knowledge of invertebrate zoology has increased over the past two centuries, structures of astonishing similarity which were first thought to be homologous were later found to be only analogous. In botany, too, homologous resemblance has often had to be later reclassified as convergence, or analogy, as knowledge has increased. Wardlaw comments that in the immediate post-Darwinian era:[21]

> Similar formal and structural characters in different species, genera and higher systematic units were accepted as being homologous. Later, as contemplation of the accumulating morphological evidence brought the realization that comparable developments were to be observed in species that could not be regarded as being closely related

genetically. This led to a recognition of the fact that parallel evolution must have been very general. The more the evidence was critically examined, the more important these parallel or homoplastic development were seen to be.

Invariably, as biological knowledge has grown, common genealogy as an explanation for similarity has tended to grow ever more tenuous. Clearly, such a trend carried to the extreme would hold calamitous consequences for evolution, as homologous resemblance is the very *raison d'être* of evolution theory. Without the phenomenon of homology – the modification of similar structures to different ends – there would be little need for a theory of descent with modification.

It turns out, then, that the problem of unity of type is not nearly as readily explicable in terms of evolution theory as is generally assumed. Darwin's jibe at Owen now seems increasingly hollow. There is still no satisfactory biological explanation for the phenomenon. Like so much of the other circumstantial "evidence" for evolution, that drawn from homology is not convincing because it entails too many anomalies, too many counter-instances, far too many phenomena which simply do not fit easily into the orthodox picture. The failure of homology to substantiate evolutionary claims has not been as widely publicised as have the problems in paleontology. Comparative embryology is a less glamorous pursuit than the biology of dinosaurs. Nonetheless, it fits into the general theme that advances in knowledge are not making it easier to reduce nature to the Darwinian Paradigm.

The discussion in the past three chapters indicates that the facts of comparative anatomy and the pattern of nature they reveal provide nothing like the overwhelming testimony to the Darwinian model of evolution that is often claimed. Simpson's claim that "the facts simply do not make sense unless evolution is true"[22] or Dobzhansky's that "nothing in biology makes sense except in the light of evolution"[23] are simply not true if by the term evolution we mean a gradual process of biological change directed by natural selection.

It is true that both genuine homologous resemblance, that is, where the phenomenon has a clear genetic and embryological basis (which as we have seen above is far less common than is often presumed), and the hierarchic patterns of class relationships are suggestive of some kind of *theory of descent*. But neither tell us anything about *how* the descent or evolution might have occurred, as to whether the process was gradual or sudden, or as to whether the causal mechanism was Darwinian, Lamarckian, vitalistic or even creationist. Such a theory of descent is therefore devoid of any significant meaning and equally compatible with almost any philosophy of nature.

In the last analysis the facts of comparative anatomy provide no evidence for evolution in the way conceived by Darwin, and even if we were to construe with the eye of faith some "evidence" in the pattern of diversity for the Darwinian model of evolution this could only be seen, at best, as indirect or circumstantial.

Readers familiar with Conan Doyle's Sherlock Holmes stories will recall the views of the great detective:[24]

> "Circumstantial evidence is a very tricky thing," answered Holmes thoughtfully; "it may seem to point very straight to one thing, but if you shift your own point of view a little, you may find it pointing in an equally uncompromising manner to something entirely different"....
> "There is nothing more deceptive than an obvious fact".

The same deep homologous resemblance which serves to link all the members of one class together into a natural group also serves to distinguish that class unambiguously from all other classes. Similarly, the same hierarchic pattern which may be explained in terms of a theory of common descent, also, by its very nature, implies the existence of deep divisions in the order of nature. The same facts of comparative anatomy which proclaim unity also proclaim division; while resemblance suggests evolution, division, especially where it appears profound, is counter-evidence against the whole notion of transmutation.

NOTES

1. Smith, H.M. (1960) *Evolution of Chordate Structure*, Holt, Rinehart and Winston, New York, from Figure 7.14, p176.
2. Darwin, C. (1872) The Origin of Species, 6th ed, 1962, Collier Books, New York pp434–5.
3. ibid, p435.
4. *Encyclopaedia Britannica* (1981) 15th ed, Encyclopaedia Britannica Inc, Chicago, Macropaedia vol7, p8.
5. Abercombie, M., Hickman, C.J. and Johnson, M. (1961) *A Dictionary of Biology*, Penguin Books Ltd, Harmondsworth, p114.
6. De Beer, G. (1971) *Homology: An Unsolved Problem*, Oxford University Press, London.

7. ibid, p 13.
8. ibid, p 8.
9. Imms, A. D. (1957) *A General Textbook of Entomology*, 9th ed, Methuen and Co, London, p 236.
10. ibid, p 240.
11. Chapman, R. F. (1969) *The Insects*, English Universities Press Ltd, p 415.
12. ibid, p 416.
13. Brook, A. J. (1964) *The Living Plant*, Edinburgh University Press, Edinburgh, p 282.
14. ibid, pp 281 and 466.
15. Darwin, op cit, p 492.
16. Mayr, E. (1970) *Populations, Species and Evolution*, Harvard University Press, Cambridge, Mass, p 93.
17. Hadorn, E. (1961) *Development Genetics and Lethal Factors*, Methuen and Co, London, from Figure 7, p 191.
18. De Beer, op cit, p 15.
19. Hardy, A. (1965) *The Living Stream*, Collins, London, p 213.
20. Carter, G. S. (1967) *Structure and Habit in Vertebrate Evolution*, Sidgwick and Jackson, London, p 493.
21. Wardlaw, C. W. (1965) *Organization and Evolution in Plants*, Longmans, Green and Co, London, pp 68–9.
22. Simpson, G.G. (1962) in foreword of reprint of 6th ed of *Origin of Species*, op cit.
23. Dobzhansky, T. (1954) in *Evolution* (1977) eds T. Dobzhansky, F.J. Ayala, G. L. Stebbins, and J. W. Valentine, W. H. Freeman and Co, San Francisco, see frontispiece.
24. Conan Doyle, A. (1928) "The Boscombe Valley Mystery" in *The Complete Sherlock Holmes*, John Murray, London, p 79.

CHAPTER 8
The Fossil Record

But as by this theory innumerable transitional forms must have existed why do we not find them embedded in countless numbers in the crust of the earth?

The overall picture of life on Earth today is so discontinuous, the gaps between the different types so obvious, that, as Steven Stanley reminds us in his recent book *Macroevolution*, if our knowledge of biology was restricted to those species presently existing on Earth,

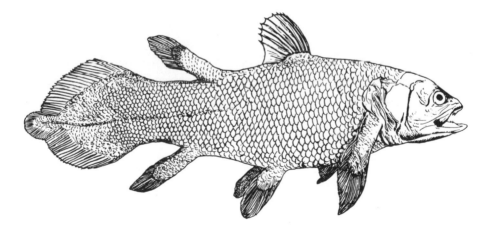

Figure 8.1: *The Coelacanth was thought to have been extinct for nearly one hundred million years until a fisherman caught a living specimen off the coast of East Africa. As a close relative of an extinct group of fishes, the Rhipidistia, considered to be the ancestors of the Amphibia and hence to all terrestrial vertebrates, its discovery provoked considerable excitement. However, examination of its soft anatomy revealed features which were not at all what was required of a close relative of the supposed ancestors of the Amphibia.* (from Gregory)[1]

"we might wonder whether the doctrine of evolution would qualify as anything more than an outrageous hypothesis."[2] Without intermediates or transitional forms to bridge the enormous gaps which separate existing species and groups of organisms, the concept of evolution could never be taken seriously as a scientific hypothesis.

The preoccupation with the problem of the intermediates, which was a feature of both popular and scholarly evolutionary thought following the publication of *The Origin of Species*, is quite understandable considering that in the 1860s and 70s evolution was still a radical idea fighting for acceptance and respectability within a biological community which was still largely typological in outlook. It was perfectly obvious to Darwin and his contemporaries, who had the difficult task of convincing their sceptical colleagues of the validity of evolution, that transitional forms were essential to the credibility of their claims. The fact that they were largely missing was acknowledged to be a major flaw in their argument.

In the decades immediately following the publication of *The Origin* it was widely believed that eventually the missing links would be found and the theory of evolution confirmed. Hence the search for them became something of an obsession; word of new fossil discoveries was greeted with considerable excitement. And it was not only among the fossils that the hope was entertained of finding the transitional forms that evolution demanded. What particularly caught the imagination of biologists, and the general public, was the prospect of finding "living links" in unexplored regions of the globe. Conan Doyle's *Lost World* captured something of the atmosphere of those times. No doubt the possibility of a vast plateau in an inaccessible part of South America surrounded by impenetrable jungle, populated by prehistoric species, a sort of living museum of the past, was somewhat melodramatic but the idea behind it was entertained not only by the public but also, although in not such a dramatic sense, by serious students of evolution. And if there were no 'lost worlds' on land then there was always the ocean, whose depths were almost entirely unexplored in 1860.

Some of the first attempts at dredging up life from the sea bed carried out in the Lofoten fjords in Norway and round the Atlantic coast of Britain in the 1860s had already revealed that certain species of sea creatures previously thought to have been extinct for millions of years were in fact still living on the ocean floor. Such discoveries encouraged the idea that the depths of the sea might be "the safest of all retreats, the secret abysses where the survivors of former geologic periods would be sure to be found."[3] And some of these survivors might be the missing links which the novel concept of evolution implied must have once existed.

When HMS *Challenger* set off in 1872 to carry out the first full-scale systematic examination of deep sea fauna there was great excitement each time a new trawl or dredge was brought up to the surface. As one of the scientific members of the expedition records "every man and boy on the ship who could possibly slip away crowded round it to see what had been fished up."[4] But as the voyage progressed and each dredge merely brought up forms already well known and related to shallow water species the excitement gradually abated. In fact, neither the *Challenger* nor any other subsequent oceanographic expedition ever dredged up any kind of missing link.

The seas and land have of course yielded many new species not known in Darwin's time as zoologists and botanists have extended their explorations into regions unexamined one century ago. A number of deep sea fish species and many invertebrates, both terrestrial and aquatic, have been discovered over the past century but all of them have been very closely related to already known groups, and in the few exceptional cases, when a quite new group of organisms has been discovered, it has invariably proved to be isolated and distinct and in no sense intermediate or ancestral in the manner required by evolution.

One of the few examples of the discovery of a quite novel type of organism occurred when a species of a hitherto unknown form of marine worm was fished up in Indonesian waters in 1900 which eventually turned out to belong to a new phylum subsequently named the Poganophora.[5] We now know that these are sessile organisms living in long stiff chitinous tubes attached to the ocean bottom. They came to public attention when the research submarine *Alvin* photographed some gigantic specimens up to two metres in length near warm water volcanic vents nine thousand meters down on the floor of the Galapagos rift. But rather than being a link between already well established phyla, the Poganophora turned out to be one of the most unusual and highly specialised types of organism ever discovered. One of their most remarkable features, which is unparalleled among any other known metazoan organisms, is the complete absence of any mouth or digestive tract and even today their mechanism of feeding is a puzzle to zoologists. Like most other invertebrate phyla, the Poganophora must be placed on an extremely distal twig of the hypothetical evolutionary tree.

However, although the possibility has always existed of finding missing links still living in unexplored parts of the globe, the major hope always lay in the direction of the fossils.

In Dawin's day only a tiny fraction of all fossil bearing strata had been examined and the number of professional paleontologists could practically be counted on two hands. Huge areas of the globe had never been explored and certainly not examined by geologists and paleontologists. Large areas of the Soviet Union, Australia, Africa and most of Asia were practically untouched. The absence of intermediates, although damaging, was not fatal in 1860, for it was reasonable to hope that many would eventually be found as geological activities increased. As Darwin rightly points out several times in the *Origin* "only a small portion of the surface of the earth has been geologically explored and no part with sufficient care" and "we continually forget how large the world is compared with the area over which our geological formations have been carefully examined."[6]

At that time the ignorance of the geology of countries outside Europe and the United States was such that it would have been as Darwin claims:[7]

> . . . about as rash to dogmatise on the succession of organic forms throughout the world, as it would for a naturalist to land for five minutes on a barren point in Australia, and then to discuss the number and range of its productions.

By stressing the very small fraction of all potentially fossil bearing strata examined in his time, Darwin was able to blunt the criticism of his opponents who found the absence of connecting links irreconcilable with organic evolution. Darwin's implication in the *Origin* that many connecting links might well be buried and awaiting discovery was also borne out to some extent by the fact that, when he was writing, new paleontological discoveries were continually being made; and this tended to confirm the idea that the picture of past life recovered from the fossils was far from complete.

Since Darwin's time the search for missing links in the fossil record has continued on an ever-increasing scale. So vast has been the expansion of paleontological activity over the past one hundred years that probably 99.9% of all paleontological work has been carried out since 1860. Only a small fraction of the hundred thousand or so fossil species[8] known today were known to Darwin. But virtually all the new fossil species discovered since Darwin's time have either been closely related to known forms or, like the Poganophoras, strange unique types of unknown affinity.

One of the most spectacular discoveries of an assemblage of new fossil species was made in 1909 by the American paleontologist Charles Doolittle Walcott when he recovered from the Burgess shale formation of British Columbia a remarkable collection of wonderfully preserved animals dating from Cambrian times, about six hundred million years ago. Alongside the many well-known forms such as jellyfish, starfish, trilobites and early molluscs present in these ancient sediments, Walcott found many species which were clearly representatives of hitherto unknown phyla and, from what has been subsequently gleaned of their biology from their fossilised remains, they would seem to be every bit as remarkable as the living Poganophora described above.

Morris and Wittington[9] recently described some of these curious forms. One of them, known aptly as *Hallucigensia*, propelled itself across the sea floor by means of seven pairs of sharply pointed, stilt-like legs. Along its back was a row of seven tentacles, each of which ended in strengthened pincers. Another unique form was *Opabinia* with five eyes across its head and a curious grasping organ that extended forward from its head and ended in a single bifurcated tip which it presumably used for catching its prey. Altogether the representatives of ten completely new invertebrate phyla were eventually recovered from the Burgess Shale, yet none of them turned out to be links between previously known phyla, only ten hitherto unknown and presumably peripheral twigs of the tree of life. As Morris and Wittington comment:[10]

> Perhaps the most intriguing problem presented by the Burgess Shale fauna is the 10 or more invertebrate genera that so far have defied all efforts to link them with known phyla. They appear to be the only known representatives of phyla whose existence had not even been suspected.

In 1947, nearly forty years after the discovery of the Burgess Shale fauna, an Australian geologist, R. C. Sprigg, made another dramatic discovery of a new assemblage of fossil forms, this time in the remote Ediacara Hills of South Australia in pre-Cambrian rocks lain down some seven hundred million years ago.[11] As with the Burgess Shale some of the species of the Ediacara fauna belonged to well known modern groups. Present in these ancient seas, for example, were corals very similar to the familiar sea pen corals which still thrive today in many of the warmer oceans. But among these familiar sea pens were other species of utterly unknown affinity, such as the strange alien form named *Tribrachidium*, a plate-like organism which possessed three equal, radiating, hooked and tentacle-fringed arms. Comments Glaesner:[12]

> Nothing like it has ever been seen among the known millions of species of animals. It recalls nothing but the three bent legs forming the coat of arms of the Isle of Man.

The discoveries in the Burgess Shale and in the Ediacaran Hills are only two examples illustrative of what has been the universal experience of paleontology, that ever since the days of the Jardin de Paris and the founding of the science in the late eighteenth century, while the rocks have continually yielded new and exciting and even bizarre forms of life, dinosaurs, ichthyosaurs and pterosaurs, in the early nineteenth century, *Hallucigensia* and *Tribrachidium* and many others in the twentieth century, what they have never yielded is any of Darwin's myriads of transitional forms. Despite the tremendous increase in geological activity in every corner of the globe and despite the discovery of many strange and hitherto unknown forms, the infinitude of connecting links has still not been discovered and the fossil record is about as discontinuous as it was when Darwin was writing the *Origin*. The intermediates have remained as elusive as ever and their absence remains, a century later, one of the most striking characteristics of the fossil record.

It is still, as it was in Darwin's day, overwhelmingly true that the first representatives of all the major classes of organisms known to biology are already highly characteristic of their class when they make their initial appearance in the fossil record. This phenomenon is particularly obvious in the case of the invertebrate fossil record. At its first appearance in the ancient paleozoic seas, invertebrate life was already divided into practically all the major groups with which we are familiar today. Not only was every major invertebrate phyla represented, but a good many of their main subgroups were also present. The molluscs, for example, the earliest representatives of the cephalopods (the group including the octopus and squid), of the bivalves (clams and oysters) or gastropods (snails and slugs), etc are all highly differentiated when they burst into the fossil record. Neither the phyla nor their main sub-divisions are linked by transitional forms. Robert Barnes summed up the current situation: ". . . the fossil record tells us almost nothing about the evolutionary origin of phyla and classes. Intermediate forms are non-existent, undiscovered, or not recognized."[13]

Curiously, the problem is compounded by the fact that the earliest representatives of most of the major invertebrate phyla appear in the fossil record over a relatively short space of geological time, about six hundred million years ago in the Cambrian era. The strata lain down over the hundreds of millions of years before the Cambrian era, which might have contained the connecting links between the major phyla, are almost completely empty of animal fossils. If transitional types between the major phyla ever existed then it is in these pre-Cambrian strata that their fossils should be found.

The story is the same for plants. Again, the first representatives of each major group appear in the fossil record already highly specialized and highly characteristic of the group to which they belong. Perhaps one of the most abrupt arrivals of any plant group in the fossil record is the appearance of the angiosperms in the era known to geologists as the Cretaceous. Like the sudden appearance of the first animal groups in the Cambrian rocks, the sudden appearance of the angiosperms is a persistent anomaly which has resisted all attempts at explanation since Darwin's time. The sudden origin of the angiosperms puzzled him. In a letter to Hooker he wrote: "Nothing is more extraordinary in the history of the Vegetable Kingdom, as it seems to me, than the *apparently* very sudden or abrupt development of the higher plants."[14] At their first appearance the angiosperms were divided into different classes, many of which have persisted with little change up to the present day. Within a space of probably less than fifty million years from their first appearance the angiosperms transformed the world's vegetation.

Again, just as in the case of the absence of pre-Cambrian fossils, no forms have ever been found in pre-Cretaceous rocks linking the angiosperms with any other group of plants. According to Daniel Axelrod[15]

> The ancestral group that gave rise to angiosperms has not yet been identified in the fossil record, and no living angiosperm points to such an ancestral alliance. In addition, the record has shed almost no light on relations between taxa at ordinal and family level.

The same pattern is true of the vertebrate fossil record. The first members of each major group appear abruptly, unlinked to other groups by transitional or intermediate forms. Already at their first appearance, although often more generalized than later representatives, they are well differentiated and already characteristic of their respective classes. Take, for example, the way the various fish groups make their appearance. In the space of less than fifty million years, starting about four hundred million years ago, a high proportion of all known fish groups appear in the fossil record. Included are many extinct fish groups such as the archaic jawless ostracoderms, the bizarre and heavily armoured placoderms as well as many representatives of modern fish forms such as the lungfish, the coelacanths and the sturgeons. The first representatives of all these groups were already so highly differentiated and isolated at their first appearance that none of them can be considered even in the remotest sense as intermediate with regard to other groups. The story is the same for the cartilaginous fish – the sharks and rays – which appear first some fifty million years later than most other fish groups. At their first appearance they too are highly specialized and quite distinct and isolated from the earlier fish groups. No fish group known to vertebrate paleontology can be classed as an ancestor of another; all are related as sister groups, never as ancestors and descendants.

The pattern repeats itself in the emergence of the amphibia. Over a period of about fifty million years, beginning about three hundred and fifty million years ago, a number of archaic and now extinct groups of amphibia make their appearance as fossils. Again, however, each group is distinct and isolated at its first appearance and no group can be construed as being the ancestor of any other amphibian group. Similarly, when the modern amphibian groups such as frogs and toads, salamanders and newts make their first appearance many millions of years later, they are again at their first occurrence differentiated and isolated from the earlier amphibian forms. The same pattern is evident as the various reptile and mammalian groups make their first appearance in the fossil record.

The overall character of the fossil record as it stands today was superbly summarized in an article by G. G. Simpson prepared for the Darwin Centenary Symposium held in Chicago in 1959. Simpson is a leading paleontologist whose testimony to the reality of the gaps in the fossil record has considerable force. As he points out, it is one of the most striking features of the fossil record that most new kinds of organisms appear abruptly:[16]

> They are not, as a rule, led up to by a sequence of almost imperceptibly changing forerunners such as Darwin believed should be usual in evolution. A great many sequences of two or a few temporally inter-grading species are known, but even at this level most species appear without known immediate ancestors, and really long, perfectly complete sequences of numerous species are exceedingly rare. Sequences of genera immediately successive or nearly so at that level (not necess-arily from one genus to the next), are more common and may be longer than known sequences of species. But the appearance of a new genus in the record is usually more abrupt than the appearance of a new species; the gaps involved are generally larger, that is, when a new genus appears in the record it is usually well separated morphologically from the most nearly similar other known genera. This phenomenon becomes more universal and more intense as the hierarchy of categories is ascended. Gaps among known species are sporadic and often small. Gaps among known orders, classes and phyla are systematic and almost always large.

In effect Simpson is admitting that the fossils provide none of the crucial transitional forms required by evolution. Moreover, because the gaps between the taxa in the fossil record increase in intensity as the taxonomic hierarchy is ascended, the incorporation of fossil taxa into the *Systema Naturae* leaves the whole orderly hierarchic scheme intact. All the new forms of life which have been uncovered by paleontology invariably relate, whether closely or distantly, as sister species to already known forms and must therefore be placed periph-erally in any hypothetical evolutionary tree.

The virtual complete absence of intermediate and ancestral forms from the fossil record is today recognised widely by many leading paleontologists as one of its most striking characteristics, so much so that those authorities who have adopted the cladistic framework now take it as axiomatic, that, in attempting to determine the relationships of fossil species, in the words of a recent British Museum publication: "we assume that none of the fossil species we are considering is the ancestor of the other."[17]

The fossils have not only failed to yield the host of transitional forms demanded by evolution theory, but because nearly all extinct species and groups revealed by paleontology are quite distinct and isolated as they burst into the record, then the number of hypothetical connecting links to join its diverse branches is necessarily greatly increased. This is particularly obvious in the case of groups like the ichthyosaurs which are intensely isolated from their most closely related sister groups. For example, a hypothetical evolutionary tree, linking lizards and turtles might have had the following form before the discovery of the ichthyosaurs:

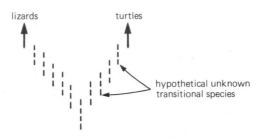

but after the discovery of the ichthyosaurs it must be redrawn thus:

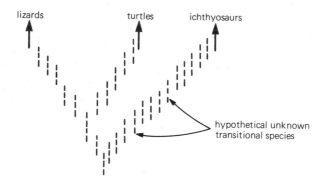

The absence of transitional forms from the fossil record is dramatically obvious (even to a non-specialist without any knowledge of comparative morphology) where a group possesses some significant skeletal specialization or adaptation which is absent in its presumed ancestral type.

Take, for example, the fish-amphibian transformation. It is generally presumed that amphibia evolved from fish and even the order of fish, the Rhipidistia, has been specified. However, transitional forms are lacking. The first amphibian had well-developed fore-and hindlimbs of normal tetrapod type which were fully capable of supporting terrestrial motion. In Figure 8.2 the first amphibian is shown alongside its nearest presumed fish ancestor. Some authorities have named near relatives of the presumed ancestors of the three major classes of vertebrates capable of flight, the pterosaurs (the flying reptiles now extinct), birds and bats, but again there is an enormous gap between the first representative of each of these three volant classes and their nearest presumed ancestral types. This can be seen again in Figure 8.2 where the first fossil known of each of the three vertebrate classes capable of flight is placed alongside the nearest terrestrial non-volant form from which each class is supposed to have evolved. The gaps between the large aquatic vertebrate groups such as ichthyosaurs, plesiosaurs, whales, seals, sea cows etc and their nearest presumed terrestrial ancestors are also dramatically obvious as can be seen in the Figure 8.2.

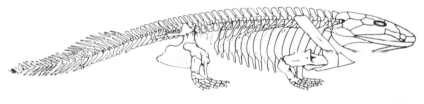

Ichthyostega, one of the earliest amphibians from the late Devonian. (from Jarvik)[18]

Eusthenopteron, (from Gregory and Raven)[19], *a representative of the Rhipidistia, the group of bony fish thought by some to be the ancestor of the amphibians, also from the late Devonian*[20, 21]

Figure 8.2a: *The earliest known amphibian alongside a Rhipidistian fish.*

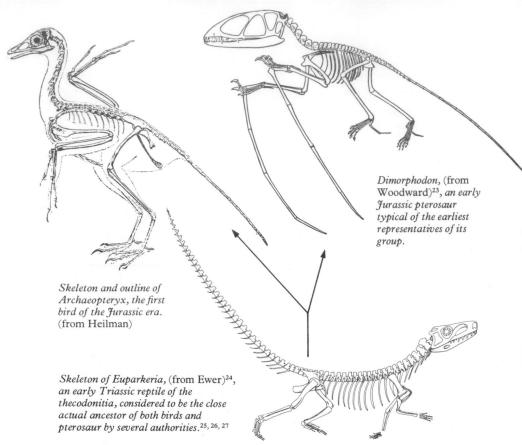

Dimorphodon, (from Woodward)[23], *an early Jurassic pterosaur typical of the earliest representatives of its group.*

Skeleton and outline of Archaeopteryx, the first bird of the Jurassic era. (from Heilman)

Skeleton of Euparkeria, (from Ewer)[24], *an early Triassic reptile of the thecodonitia, considered to be the close actual ancestor of both birds and pterosaur by several authorities.*[25, 26, 27]

Figure 8.2b: *The first bird, an early Pterosaur and their closest non-volant relative.*

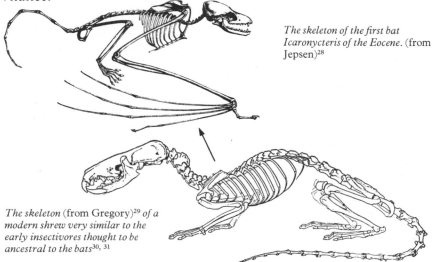

The skeleton of the first bat Icaronycteris of the Eocene. (from Jepsen)[28]

The skeleton (from Gregory)[29] *of a modern shrew very similar to the early insectivores thought to be ancestral to the bats*[30, 31]

Figure 8.2c: *The skeleton of the first bat alongside that of a small non-volant mammal.*

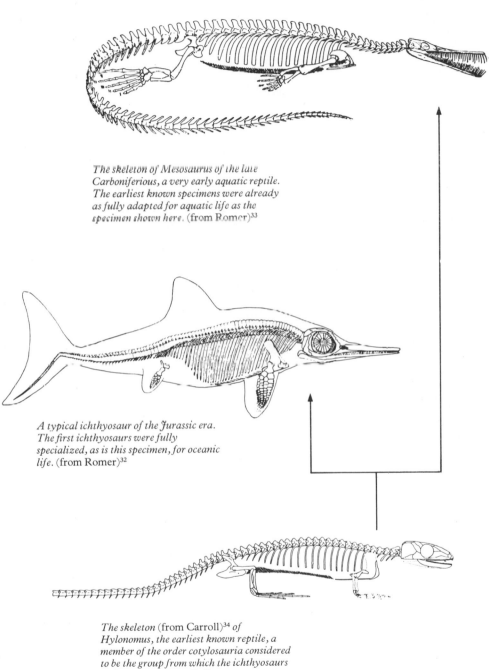

The skeleton of Mesosaurus of the late Carboniferous, a very early aquatic reptile. The earliest known specimens were already as fully adapted for aquatic life as the specimen shown here. (from Romer)[33]

A typical ichthyosaur of the Jurassic era. The first ichthyosaurs were fully specialized, as is this specimen, for oceanic life. (from Romer)[32]

The skeleton (from Carroll)[34] of Hylonomus, the earliest known reptile, a member of the order cotylosauria considered to be the group from which the ichthyosaurs and mesosaurs arose.[35, 36]

Figure 8.2d: *An Ichthyosaur and another type of aquatic reptile, Mesosaurus, and one of their closest known terrestrial relatives.*

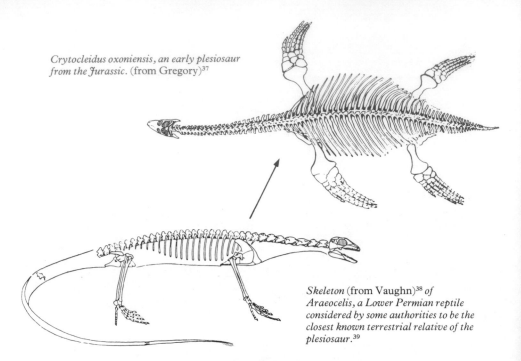

Crytocleidus oxoniensis, an early plesiosaur from the Jurassic. (from Gregory)[37]

Skeleton (from Vaughn)[38] *of Araeocelis, a Lower Permian reptile considered by some authorities to be the closest known terrestrial relative of the plesiosaur.*[39]

Figure 8.2e: *A typical early Plesiosaur and its nearest known terrestrial relative.*

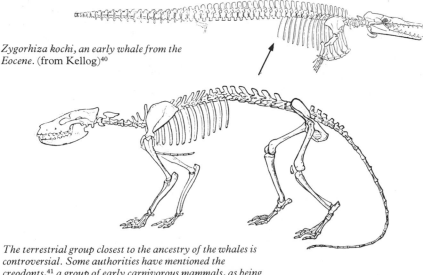

Zygorhiza kochi, an early whale from the Eocene. (from Kellog)[40]

The terrestrial group closest to the ancestry of the whales is controversial. Some authorities have mentioned the creodonts,[41] *a group of early carnivorous mammals, as being possibly ancestral to the whales. A typical example of a creodont is Sinopa, shown here from the lower Eocene.*[42]

Figure 8.2f: *An early whale and one of its nearest terrestrial relatives.*

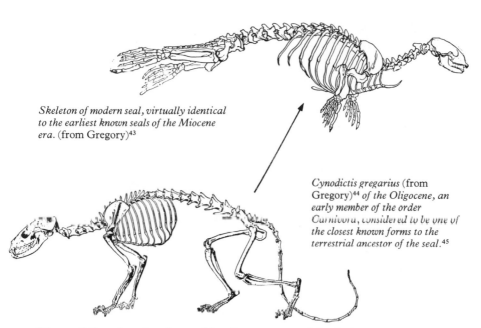

Skeleton of modern seal, virtually identical to the earliest known seals of the Miocene era. (from Gregory)[43]

Cynodictis gregarius (from Gregory)[44] of the Oligocene, an early member of the order Carnivora, considered to be one of the closest known forms to the terrestrial ancestor of the seal.[45]

Figure 8.2g: *A seal and one of its closest terrestrial relatives.*

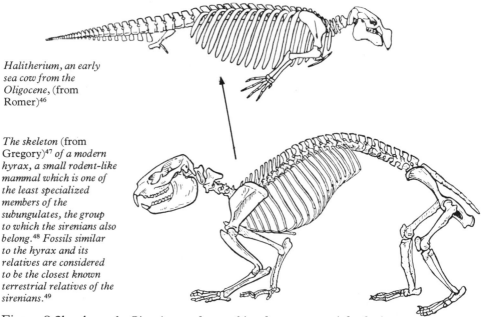

Halitherium, an early sea cow from the Oligocene, (from Romer)[46]

The skeleton (from Gregory)[47] of a modern hyrax, a small rodent-like mammal which is one of the least specialized members of the subungulates, the group to which the sirenians also belong.[48] Fossils similar to the hyrax and its relatives are considered to be the closest known terrestrial relatives of the sirenians.[49]

Figure 8.2h: *An early Sirenian and one of its closest terrestrial relatives.*

It would be pointless to continue citing examples to illustrate the discontinuous nature of the fossil record. Anyone who doubts the reality of the gaps may either take the word of leading paleontologists or simply open one of the standard works on paleontology such as Romer's *Vertebrate Paleontology*[50] or Schrock and Twenhofel's *Invertebrate Paleontology*[51] and examine any of the stratigraphic charts showing the abundance of various groups during different geological eras and dotted lines suggesting their hypothetical phylogenetic relationships (see Figure 8.3). Even a cursory glance shows clearly that profound and undoubted discontinuities do in fact exist.

There is no doubt that as it stands today the fossil record provides a tremendous challenge to the notion of organic evolution, because to close the very considerable gaps which at present separate the known groups would necessarily have required great numbers of transitional forms. Over and over again in the *Origin* Darwin reiterates the same point, leaving the reader in no doubt as to his belief that to bridge the gaps innumerable transitional forms would have to be postulated:[52]

> By the theory of natural selection all living species have been connected with the parent-species of each genus, by differences not greater than we see between the natural and domestic varieties of the same species at the present day; and these parent-species, now generally extinct, have in their turn been similarly connected with more ancient forms; and so on backwards, always converging to the common ancestor of each great class. So that the number of intermediate and transitional links, between all living and extinct species, must have been *inconceivably* great. But assuredly, if this theory be true, such have lived upon the earth.
>
> [*emphasis added*]

Darwin's insistence that gradual evolution by natural selection would require inconceivable numbers of transitional forms may have been something of an exaggeration but it is hard to escape concluding that in some cases he may not have been so far from the mark. Take the case of the gap between modern whales and land mammals. All known aquatic or semi-aquatic mammals such as seals, sea cows (sirenians) or otters are specialized representatives of distinct orders and none can possibly be ancestral to the present-day whales. To bridge the gap we are forced therefore to postulate a large number of entirely extinct hypothetical species starting from a small, relatively unspecialized land mammal like a shrew and leading successively

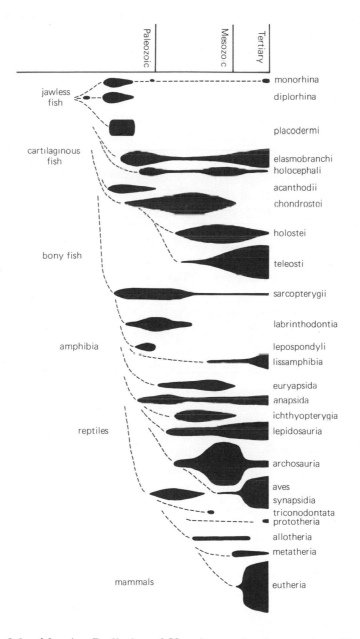

Figure 8.3: *Adaptive Radiation of Vertebrates showing stratigraphic abundance of the major vertebrate groups through time. The dotted lines represent hypothetical lineages required by evolution to link the various groups together.* (from Romer and Carter)[53, 54]

through an otter-like stage, seal-like stage, sirenian-like stage and finally to a putative organism which could serve as the ancestor of the modern whales. Even from the hypothetical whale ancestor stage we need to postulate many hypothetical primitive whales to bridge the not inconsiderable gaps which separate the modern filter feeders (the baleen whales) and the toothed whales. Moreover, it is impossible to accept that such a hypothetical sequence of species which led directly from the unspecialized terrestrial ancestral form gave rise to no collateral branches. Such an assumption would be purely *ad hoc*, and would also be tantamount to postulating an external unknown directive influence in evolution which would be quite foreign to the spirit of Darwinian theory and defeat its major purpose of attempting to provide a natural explanation for evolution. Rather, we must suppose the existence of innumerable collateral branches leading to many unknown types. This was clearly Darwin's view and it implies that the total number of species which must have existed between the discontinuities must have been much greater than the number of species on the shortest direct evolutionary pathway. In the diagram opposite, which shows a hypothetical lineage leading from a land mammal to a whale, while there are ten hypothetical species on the direct path, there are an additional fifty-three hypothetical species on collateral branches.

Considering how trivial the differences in morphology usually are between well-defined species today, such as rat-mouse, fox-dog, and taking into account all the modifications necessary to convert a land mammal into a whale – forelimb modifications, the evolution of tail flukes, the streamlining, reduction of hindlimbs, modifications of skull to bring nostrils to the top of head, modification of trachea, modifications of behaviour patterns, specialized nipples so that the young could feed underwater (a complete list would be enormous) – one is inclined to think in terms of possibly hundreds, even thousands, of transitional species on the most direct path between a hypothetical land ancestor and the common ancestor of modern whales.

Further, when we repeat the above process to envisage the bridging of all the gaps between different types of organisms and to connect all the unique and isolated groups such as whales, icthyosaurs, pleisiosaurs, turtles, seals and sea cows we are forced to admit with Darwin that in terms of gradual evolution, considering all the collateral branches that must have existed in the crossing of such gaps, the number of transitional species must have been inconceivably great.

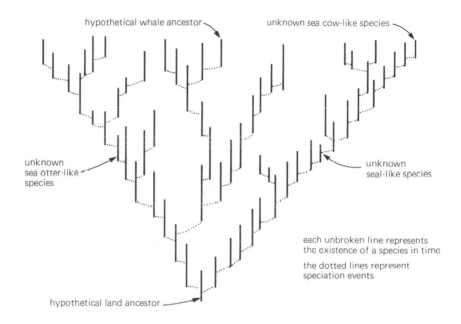

hypothetical whale ancestor

unknown sea cow-like species

unknown
sea otter-like
species

unknown
seal-like species

each unbroken line represents
the existence of a species in time

the dotted lines represent
speciation events

hypothetical land ancestor

Despite the generally discontinuous character of the fossil record there are some exceptional cases where a species does appear to be intermediate with respect to other groups. The classic case of this is, of course, *Archaeopteryx*, a picture of which is shown in Figure 8.2. This primitive bird did indeed possess certain skeletal reptilian features – teeth, a long tail, claws on its wings. However, in one respect, flight, the most characteristic feature of birds, *Archaeopteryx* was already truly bird. On its wing there were flight feathers as fully developed as any modern bird, and recent research reported in 1979[55] suggests that it was as capable of powered flight as a modern bird.

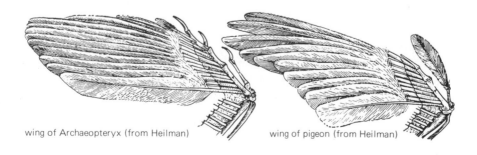

wing of Archaeopteryx (from Heilman)   wing of pigeon (from Heilman)

Commenting on the structure of the wing of *Archaeopteryx*, Savile remarks:[57]

> The eight primaries are graduated proximally and distally to form a nearly elliptical wing tip of surprisingly modern appearance. . . . The wings were not preserved in fully extended position, but they seem to have been nearly elliptical in outline, and, except for the lack of an alula or emarginated primaries, are not unlike those of many passerines in general appearance.

No doubt it can be argued that *Archaeopteryx* hints of a reptilian ancestry but surely hints do not provide a sufficient basis upon which to secure the concept of the continuity of nature. Moreover, there is no question that this archaic bird is not led up by a series of transitional forms from an ordinary terrestrial reptile through a number of gliding types with increasingly developed feathers until the avian condition is reached. A much more convincing intermediate would be something like Heilman's imaginary "pro-avis", a supposed hypothetical ancestor of *Archaeopteryx* which glided through the trees assisted by partially developed feathers (see Figure 9.1).

Another group designated intermediate is a group of reptile-like amphibians, one representative of which, *Seymouria*, is described by Romer:[58]

> a modest form from the lower Permian exhibits such a combination of amphibian and reptilian characters that its proper position in the classification of the vertebrates has been much disputed. . . . *Seymouria* thus seems . . . *almost exactly on the dividing line* between amphibians and reptiles.
>
> [*emphasis added*]

In terms of purely skeletal characteristics *Seymouria* would appear to be a convincing intermediate, but there is a serious drawback. The major difference between amphibians and reptiles lies in their reproductive systems. Amphibians lay their eggs in water and their larvae undergo a complex metamorphosis (like a tadpole) before reaching the adult stage. Reptiles develop inside a hard shell-encased egg and are perfect replicas of the adult on first emerging. The problems of envisaging the gradual evolution of the reptilian egg are dealt with later, but the point at issue here is that skeletal characteristics alone are insufficient for designating a particular organism or species as intermediate. Recently a fossil of an immature form closely related to *Seymouria* has been found bearing laval gills (like a tadpole) which suggests that this group of amphibians were wholly amphibian in their reproductive system.[59] There is a further difficulty with *Seymouria* and that is that it appears rather too late in the fossil record to be an ancestor of the reptiles.

To demonstrate that the great divisions of nature were really bridged by transitional forms in the past, it is not sufficient to find in the fossil record one or two types of organisms of doubtful affinity which might be placed on skeletal grounds in a relatively intermediate position between other groups. The systematic status and biological affinity of a fossil organism is far more difficult to establish than in the case of a living form, and can never be established with any degree of certainty. To begin with, ninety-nine per cent of the biology of any organism resides in its soft anatomy, which is inaccessible in a fossil. Supposing, for example, that all marsupials were extinct and the whole group was known only by skeletal remains – would anyone guess that their reproductive biology was so utterly different from that of placental mammals and in some ways even more complex?

Modern birds differ greatly from reptiles in many physiological and anatomical characteristics, particularly, for example, in their central nervous, cardiovascular and respiratory systems (the respiratory system of birds is discussed in detail in the next chapter) but, because information about the soft biology of a fossil form is difficult to obtain from its skeletal remains, to what extent *Archaeopteryx* was avian in its major organ systems will always be largely a matter of conjecture.

One aspect of an organism's soft biology which can sometimes be studied in a fossil is the gross morphology of the brain. This can be done by preparing a cranial endocast of the intracranial cavity in the skull which reveals the gross shape and outline of the brain.[60] On the evidence available from study of the cranial endocast of *Archaeopteryx*, it would seem that its brain was essentially avian in all important respects, exhibiting typical avian cerebral hemispheres and cerebellum[61] (the part of the brain involved in balance and the coordination of fine motor activities), a part of the brain proportionally larger in birds than in any other class of vertebrates and generally considered to be an adaptation necessary for the control of the highly complex motor activities involved in powered flight. The possession of an essentially avian central nervous system lends further support to the idea, based on the basically modern form of its flight feathers and wing, that *Archaeopteryx* was as capable of powered flight as a typical modern bird. If *Archaeopteryx* was indeed capable of powered flight, might it not also have possessed, of necessity, a fully avian heart, circulatory and respiratory system to supply the vastly increased demand for oyxgen that occurs during powered flight? In other words, might it not have been as avian as any other bird in all important anatomical and physiological characteristics?

Then there is the problem of convergence. Nature abounds in examples of convergence: the similarity in overall shape of whales, ichthyosaurs and fishes; the similarity in the bone structure of the flippers of a whale and an ichthyosaur; the similarity of the forelimbs of a mole and those of the insect, the molecricket; the great similarity in the design of the eye in vertebrates and cephalopods and the profound parallelism between the cochlea in birds and mammals. In all the above cases the similarities, although very striking, *do not imply any close biological relationship.*

A fascinating example of convergence is the similarity between the placental and marsupial dogs. The dog-like carnivore, the thylacine, known locally in Australia as the Tasmanian wolf, lived until recently in the remote rain forests of southwest Tasmania. Although as a marsupial the thylacine was quite unrelated to the placental dog, it was incredibly similar in gross appearance and in skeletal structure, teeth, skull, etc, so similar in fact that only a skilled zoologist could distinguish them.

Anyone who has been privileged to handle, as I have, both a marsupial and placental dog skull will attest to the almost eerie degree of convergence between the thylacine and the placental dog. Yet in terms of the soft anatomy of their reproductive systems, there is an enormous difference between the two groups.

A particularly interesting case which illustrates both the problem of convergence and the danger of judging overall biology on skeletal grounds is that of the rhipidistian fishes. For nearly a century these ancient lobe finned fishes, as they are often known, have been generally considered to be ideal amphibian ancestors and have been classed as intermediate between fish and the terrestrial vertebrates. This judgment was based on a number of skeletal features including the pattern of their skull bones, the structure of their teeth and vertebral columns and even the pattern of bones in their fins, in all of which they closely resembled the earliest known amphibians. It was assumed that their soft biology would be also transitional between that of typical fish and amphibia.

But in 1938 fishermen in the Indian Ocean, off Cape Province in South Africa, hauled to the surface a living relative of the ancient Rhipidistia – the coelacanth. It was an astonishing discovery, as the coelacanth had been thought to be extinct for a hundred million years. Because the coelacanth is a close relative of the Rhipidistia, here at last was the opportunity to examine first hand the biology of one of the classic evolutionary links. Its discovery provoked considerable excitement. Peter Forey comments:[63]

We had to wait nearly one hundred years before discovery of the Recent coelacanth. During that time many fossil coelacanths were described and, on the basis of osteological features, their systematic position as near relatives of the extinct rhipidistians and as tetrapod cousins had become part of "evolutionary fact", perpetuated today in textbooks. Great things were therefore expected from the study of the soft anatomy and physiology of *Latimeria*. With due allowance for the fact that *Latimeria* is a truly marine fish, it was expected that some insight might be gained into the soft anatomy and physiology of that most cherished group, the rhipidistians. Here, at last, was a chance to glimpse the workings of a tetrapod ancestor. These expectations were founded on two premises. First, that rhipidistians are the nearest relatives of tetrapods and secondly, that *Latimeria* is a rhipidistian derivative.

But examination of the living coelacanth proved very disappointing. Much of its soft anatomy, particularly that of the heart, intestine and brain, was not what was expected of a tetrapod ancestor. As Barbara Stahl writes:[64]

the modern coelacanth shows no evidence of having internal organs preadapted for use in a terrestrial environment. The outpocketing of the gut that serves as a lung in land animals is present but vestigial in Latimeria. The vein that drains its wall returns blood not to the left side of the heart as it does in all tetrapods but to the sinus venosus at the back of the heart as it does directly or indirectly in all osteichthyans except lungfishes. The heart is characteristically fish-like in showing no sign of division into left and right sides, and the gut, with its spiral-valved intestine, is of a type common to all fishes except the most advanced ray-fins.

Clearly, if the soft biology of the rhipidistian fish resembled to any extent their coelacanth cousins then, however great their similarity to the earliest amphibia in certain skeletal features, in terms of their overall biology they must have been far removed. Of course, it is always possible to argue that the skeletal similarities between the coelacanth and the Rhipidistia are only convergence and hence superficial but if the argument is accepted in this case can it be refused in the case of the similarities between the rhipidistian fish and the amphibia?

If the case of the coelacanth illustrates anything, it shows how difficult it is to draw conclusions about the overall biology of organisms from their skeletal remains alone. Because the soft biology of extinct groups can never be known with any certainty then obviously the status of even the most convincing intermediates is bound to be insecure. The coelacanth represents yet another instance where a newly discovered species, which might have provided the elusive evidence of intermediacy so long sought by evolutionary biology, ultimately proved to be only another peripheral twig on the presumed tree of life.

There is now a body of opinion[65] which views the lungfish to be far closer to the amphibia than the Rhipidistia, a view that was common in the nineteenth century, and considers the skeletal similarities between rhipidistian fish and amphibia as only convergence.

The fact that several quite separate rhipidistian groups all exhibited, to different degrees, amphibian-like skeletal characteristics means that in most rhipidistians the similarity was indeed convergence as only one group could possibly be the ancestor of the amphibian.

Similar considerations cloud the status of the other classic intermediate groups such as the mammal-like reptiles, a group of extinct reptiles in which the morphology of the skull and jaw was very close to the mammalian condition. The possibility that the mammal-like reptiles were completely reptilian in terms of their anatomy and physiology cannot be excluded. The only evidence we have regarding their soft biology is their cranial endocasts and these suggest that, as far as their central nervous systems were concerned, they were entirely reptilian. Jerison, who has probably had more experience studying the cranial endocasts of fossil species than any other authority in this field, comments on the mammal-like reptile brains:[66]

. . . these animals had brains of typical lower vertebrate size . . . since their endocasts were all very near the volume of these expected brain sizes and since the endocasts present maximum limits on their brain sizes, the mammal-like reptiles could not have had brains that approached a mammalian size. . . . The mammal-like reptiles, in short, were reptilian and not mammalian with respect to the evolution of their brains. . . . There are few suggestions of mammalian features in the brains of the mammal-like reptiles. . . . The forebrain, to the extent that its position is identifiable, was of reptilian size and shape. This was not the case in the earliest known fossil mammal. . . . The earliest mammal for which there is reasonable evidence, *Triconodon* of the Upper Jurassic period, was apparently already at or near the level of living "primitive" mammals such as the insectivores or the Virginia opossum. It was certainly larger brained than its reptilian ancestors of comparable body size.

Moreover, many quite separate groups of mammal-like reptiles exhibited skeletal mammalian characteristics, yet only one group can have been the hypothetical ancestor of the mammals. Again, as with the rhipidistian fishes, the similarities must have been in most cases merely convergence.

This possibility was raised in the 1930s by several continental zoologists whose views have been largely forgotten in the English speaking world. Rensch refers to some of these authorities:[67]

Beurlen (1937) regarded them as a manifestation of a "typical systemic regularity" (Gestaltgesetzlichkeit) which cannot be interpreted on the basis of mutation and selection. Dacque (1935) pointed out that in some cases parallel developments occurred even in heterogeneous groups during certain geological periods, and he coined the term "time signatures". By this statement Dacque tried to interpret the phylogenetic links, such as the theromorphs, which are intermediate between the reptiles and the mammals, *as having only accidental resemblance to the mammals and not being ancestral to them.* Consequently, he doubted the phylogenetic relations of the more strongly differing structural types.

[emphasis added]

Given the tremendous diversity of life and the ubiquity of the phenomenon of convergence, it is bound to be the case that certain fossil organisms which appear to be very close on skeletal grounds were in fact in terms of their overall biology only distantly related, like the placental and marsupial dogs. Further, there is always the possibility that groups, such as the mammal-like reptiles which have left no living representative, might have possessed features in their soft biology completely different from any known reptile or mammal which would eliminate them completely as potential mammalian ancestors, just as the discovery of the living coelacanth revealed features in its soft anatomy which were unexpected and cast doubt on the ancestral status of its rhipidistian relatives.

It is clear that there are formidable problems in interpreting evidence for continuity on the basis of skeletal remains. Consequently if the fossil record is to provide any grounds for believing that the great divisions of nature are not the unbridgeable discontinuities postulated by Cuvier, it is not sufficient that two groups merely approach one another closely in terms of their skeletal morphology. The very least required would be an unambiguous continuum of transitional species exhibiting a perfect gradation of skeletal form leading unarguably from one type to another. But the fact is that, as Stanley put it:[69]

> The known fossil record fails to document a *single example of phyletic (gradual) evolution accomplishing a major morphologic transition* and hence offers no evidence that the gradualistic model can be valid.
>
> [*emphasis added*]

The only sort of evolution documented in the fossil record are several instances where a relatively minor morphological transformation can be traced through a convincing series of fossil forms. The best known case is probably that of the horse, which starts with the original dog-sized horse, *Eohippus*, which lived about sixty million years ago and leads gradually to the modern horse of today (see Figure 8.4).

The horse series is not as perfect as is commonly assumed. As Simpson points out, the single line of gradual transformation from *Eohippus* to *Equus* presented in most recent texts of evolutionary biology is largely aprocryphal. On the contrary, most of the morphological characteristics of the feet, skull and teeth, which are traditionally supposed to have exhibited an almost perfect sequence of change throughout the Tertiary, "progress from one stable adaptive level to another by a sequence of short steplike transitions" and some of the transitions are not represented in the fossil record. Simpson comments:[70]

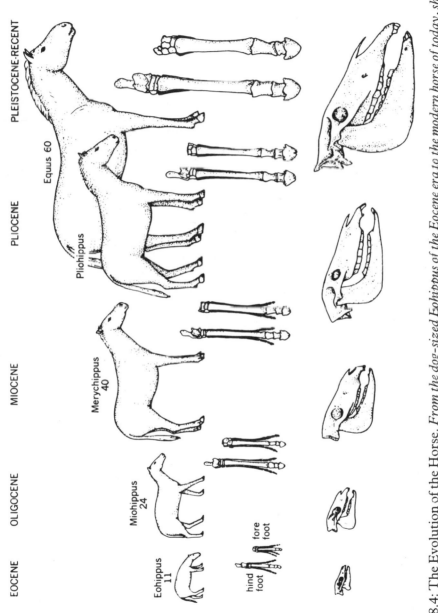

Figure 8.4: The Evolution of the Horse. *From the dog-sized Eohippus of the Eocene era to the modern horse of today, showing the changes in limb and skull structure.* (from Storer, Usinger, Mybakken and Stebbins)[68]

The most famous of all equid trends, "gradual reduction of the side toes" is flatly fictitious. There was no such trend in any line of Equidae. . . . Eocene horses all had digitigrade padded, doglike feet with four functional toes in front and three behind. In a rapid transition (not actually represented by fossils), early Oligocene horses lost one functional front toe and concentrated weight a little more on the middle hoof as a step-off point. . . . This type persisted without essential change in all browsing horses.

Even though the horse series is not exactly perfect it is difficult to avoid concluding that it probably represents a natural evolutionary sequence. But how different was *Eohippus* from a modern horse? Could such sequences be an extension of microevolution? Richard Owen, the leading British comparative anatomist of his day, wrote:[71]

A modern horse occasionally comes into the world with the supplementary ancestral hoofs. From Valerius Maximus, who attributes the variety to *Bucephalus*, downwards, such "polydactyle" horses have been noted as monsters and marvels. In one of the latest examples, the inner splint-bone, answering to the second metacarpal of the pentadactyle foot, supported phalanges and a terminal hoof, in position and proportion to the middle hoof, resembling the corresponding one in *Hipparion*.

In relation to actual horses such specimens figure as "monstra per excessum"; but, in relation to miocene horses, they would be normal, and those of the present day would exemplify "monstra per defectum." The mother of a "monstrous" tridactyle colt might repeat the anomaly and bring forth a tridactyle "filly"; just as, at San Salvador, the parents of a family of six had two of the series born with defective brain and of dwarf size: they were "male" and "female"; and these strange little idiots are exhibited as "Aztecs". The pairing of the horses with the metapodials bearing, according to type, phalanges and hoofs, might restore the race of hipparions.

Many of the other well-known series often depicted in textbooks are not nearly as convincing as the horse when subjected to detailed analysis. The elephant "sequence", for example, which starts with a hypothetical ancestral form presumed to have been somewhat similar to the modern *Hyrax* (a small rodent-like animal found today in Africa, see Figure 8.2) and passes, via the pig-sized *Moeritherium* of the late Eocene, through a series of forms which show an increasing approach in trunk size, teeth morphology etc to the modern elephant, fails to convince Sylvia Sikes:[72]

> . . . it requires extreme elasticity of the imagination to see anything more than a very superficial resemblance between the available parts of the skeletons of the earliest hyraces and those of the *Proboscidea*. . . in the light of recent comparative studies on the anatomy, physiology, ecology and ethology of the living members of these orders, it is apparent that in the past disproportionate weight was sometimes given to skeletal affinities, while other important characteristics were over-looked.

She continues:

> . . . perhaps we should admit that the siting of *Moeritherium* in an intermediate position in the family tree savours more of the artistic requirements of the drawing board than of an honest admission of ignorance as to its proper position.

Considering that the total number of known fossil species is nearly one hundred thousand, the fact that the only relatively convincing morphological sequences are a handful of cases like the horse, which do not involve a great deal of change, and which in many cases like the elephant may not even represent phylogenetic sequences at all, serves to emphasize the remarkable lack of any direct evidence for major evolutionary transformations in the fossil record.

A great deal has been made of the horse series and other similar cases. The traditional view is that they provide powerful evidence of the reality of evolution; and that what has happened in the case of the horse happened in all other cases, but the fossil links were not preserved or have not yet been discovered. In other words, the horse is the exception which proves the rule.

It is possible to view such series in a very different light and read the fossil evidence directly as it stands; and infer that what is exceptional about such sequences is not their preservation but rather the fact that they occurred. They may be exceptions which prove a very different rule: that in general, nature cannot be arranged in terms of sequences and that where sequence does exist it is exceptional or relatively trivial.

Moreover, there is another aspect of horse evolution which casts a shadow over its usefulness as the example *par excellence* of gradual evolutionary transformations. The difference between *Eohippus* and the modern horse is relatively trivial, yet the two forms are separated by sixty million years and at least ten genera and a great number of species. The horse series therefore tends to emphasize just how vast must have been the number of genera and species if all the diverse forms of life on Earth had really evolved in the gradual way that Darwinian evolution implies. If the horse series is anything to go by their numbers must have been indeed the "infinitude" that Darwin imagined. If ten genera separate *Eohippus* from the modern horse then think of the uncountable myriads there must have been linking such diverse forms as land mammals and whales or molluscs and arthropods. Yet all these myriads of life forms have vanished mysteriously, without leaving so much as a trace of their existence in the fossil record.

Basically, three explanations have been put forward to explain the gaps in the fossil record: Firstly, *insufficient search*, ie that not all fossil bearing strata have been examined. Secondly, the *imperfection of the record*, ie that only a fraction of the species that lived in the past have left fossil remains. Thirdly, *saltational evolution*, ie that the gaps are real and that evolution occurred in a series of jumps.

The hope of uncovering the missing links in unexplored rocks is not completely dead, as is witnessed by the recent excitement which accompanied the discovery of a number of new species from previously unexamined sites in Antarctica. But the hope has greatly diminished. As Norman Newell, past curator of historical geology at the American Museum of Natural History, puts it:[73]

> . . . experience shows that the gaps which separate the highest categories may never be bridged in the fossil record. Many of the discontinuities tend to be more and more emphasized with increased collecting.

It is particularly difficult to accept insufficient search as an explanation for the gaps between the major invertebrate phyla. As we have seen, all the main invertebrate types appear already clearly differentiated very abruptly in early Cambrian rocks. An enormous effort has been made over the past century to find missing links in these rocks which might bridge the deep divisions in the animal kingdom. Yet no links have ever been found and the relationships of the major groups are as enigmatic today as one hundred years ago.

In Darwin's time no fossils of any sort were known from rocks dated before six hundred million years ago, but since then fossils of unicellular and bacterial species have been found in rocks dating back thousands of millions of years before the Cambrian era. Also, several new types of organisms which were not known one hundred years ago have been discovered in the Burgess Shale and at Ediacara, in rocks of Cambrian and late pre-Cambrian age: however, none of these discoveries have thrown any light on the origin or relationships of the major animal phyla. As we have seen, newly discovered hitherto unknown groups, whether living or fossilized, invariably prove to be distinct and isolated and can in no way be construed as connecting links in the sense required by evolution theory. James Valentine, of the University of California, comments:[74]

> There are numbers of both soft-bodied and skeletonized fossils of late Precambrian or Cambrian ages which are not assignable to living phyla, nor can any of them yet be considered as ancestral to any specific living phylum.

Over the past century a host of rationalizations have been attempted to explain the mystifying absence of primitive transitional forms in the pre-Cambrian rocks. One explanation was that before the Cambrian age aquatic animals did not possess shells or hard integuments or that they possessed shell-like structures so fragile that no traces have been found because the calcium content of the sea was supposed to have been too low to permit the secretion of calcareous shells. Another explanation was that there was a long lapse of time – "the Lipalian gap" – between the latest pre-Cambrian rocks and the earliest Cambrian rocks during which time no fossil bearing rocks were lain down. Yet another proposal was that there was an absence of intertidal and near shore littoral sediments lain down in pre-Cambrian times. Some have even claimed that all fossil bearing pre-Cambrian rocks have been denuded or metamorphosed to such an extent that their fine structure has been lost, a view no longer tenable now that microfossils of bacteria and unicellular organisms have been found throughout the pre-Cambrian rocks.

Commenting on the various hypotheses put forward to account for the lack of pre-Cambrian fossils Preston Cloud, a specialist in this field, wrote:[75]

> To such hypotheses I will comment only that (1) the availability or lack of $CaCO_3$ is not the explanation for distributions of fossils observed

– carbonate rocks are abundant, both above and below the Paleozoic–Precambrian boundary. (2) The "Lipalian" gap does not exist: sequences of sedimentary rock transitional from Precambrian into Cambrian appear to be present in Australia, the southern Great Basin, perhaps British Columbia, perhaps Arctic Canada, the Appalachian region, the eastern Baltic, Siberia, and possibly Africa. (3) An explanation that calls on the absence of Precambrian littoral and intertidal sediments is invalid; such deposits, as well as deeper water sediments, are well represented in the Precambrian, although their proportions vary. (4) Metamorphism and loss of fine structure in the Precambrian does not provide an acceptable general explanation; delicate sedimentary structures are preserved in Precambrian sediments of a variety of ages and depths of deposition with a degree of fidelity that would assure the preservation somewhere of tracks, trails, burrows, or after-death impressions of pelagic metazoan organisms if they had been present. (5) A very long interval of preskeletal metazoan evolution seems unlikely although 50 to 100 MY of mainly preskeletal development represented by the Ediacaran and approximate equivalents is not out of question. One should expect to find after-death imprints of such organisms even if not tracks or burrows and some organisms, such as brachiopods, could hardly exist except in context with a shell (Cloud, 1948). (6) Insufficient search is an increasingly unlikely explanation, although always a possible one; exhaustive searches have been made in favourable sediments. To this last point I may mention the heroic labors of Walcott through years of search by himself and employed collectors, and the phenomenal industry reported by David and Tillyard (1936) who had some fifty-five tons of hard quartzite quarried free and seven tons of selected blocks carefully split into thin slabs at one site. I myself, and others, in recent years have also searched, so far in vain, for unequivocal metazoan fossils through thousands of feet of sediments at scores of selected localities.

Considering the tremendous effort that has been made over the past century to find links between the major animal groups in the pre-Cambrian strata, one wonders what Darwin's own verdict on evolution might have been today when he was prepared to go as far in the *Origin* to admit that:[76]

> the case at present must remain inexplicable and may be truly argued as a valid argument against the views here entertained.

The second explanation, the imperfection of the record, has always been the most popular explanation for the gaps. It was also Darwin's:

"The explanation lies, as I believe, in the extreme imperfection of the geological record."[77]

However, the question as to the actual degree of imperfection of the fossil record is very controversial. There is certainly evidence that it is far from perfect, as Thomas Schopf commented recently in *Paleobiology*:[78]

> Note the case of the Order Multituberculata, the longest lived mammalian order. It is considered to range from the middle Jurassic to the end of the Eocene, 160 m.y. duration. On a stage by stage basis, fossils of this order are known to occur in stages whose cumulative duration is only 87 m.y., just 54 per cent of the duration of the order. That is, 46 per cent of the time the Multituberculata existed, there has not yet been discovered a record of the order anywhere in the world. This simply underscores the vagaries of preservation and fossilization.

On the other hand, the fact that, when estimates are made of the percentage of living forms found as fossils, the percentage turns out to be surprisingly high, suggesting that the fossil record may not be as bad as is often maintained. Of the 329 living families of terrestrial vertebrates 261 or 79.1% have been found as fossils and, when birds (which are poorly fossilized) are excluded, the percentage rises to 87.8% (see Figure 8.5).

G. G. Simpson recently estimated the percentage of living species recovered as fossils in one region of North America and concluded that, at least for larger terrestrial forms, the record may be almost complete![79] In another approach he compared the number of living genera of various categories such as insectivores, carnivores, etc in a particular region with the numbers of fossil genera of the same categories in a region of similar ecological make-up in the past. Two such ecological regions are recent Portuguese East Africa and Middle Oligocene Dakota.[80] After comparing the composition of these two faunas, Simpson concludes:[81]

> These comparisons and some other considerations suggest that surely half and probably two-thirds or more of the Middle Oligocene genera are known and that those not yet known are mainly carnivores (individually much less abundant than herbivores) and very small mammals (with less recoverability than large mammals by previous collecting methods).

Number of living orders of terrestrial vertebrates	43
Number of living orders of terrestrial vertebrates found as fossils	42
Percentage fossilised	97.7%
Number of living families of terrestrial vertebrates	329
Number of living families of terrestrial vertebrates found as fossils	261
Percentage fossilised	79.1%
Number of living families of terrestrial vertebrates excluding birds	178
Number of living families of terrestrial vertebrates found as fossils excluding birds	156
Percentage fossilised	87.8%

Figure 8.5: The Adequacy of the Fossil Record. *The table shows the percentage of living orders and living families which have been recovered as fossils.* (from Romer)[82]

According to an article by Wyatt Durham in the *Journal of Palae-ontology*,[83] as many as two per cent of all marine invertebrate species with hard skeletal components that have ever existed may be known as fossils. Assuming ten to twenty species per genus, this means that for certain groups, such as molluscs which are ideal fossil material, the percentage of genera known could be as high as fifty per cent. There are, therefore, grounds for believing that in the case of some groups appealing to the imperfection of the fossil record as an explanation for the gaps is not a particularly convincing strategy.

It is significant in this respect that many professional paleontologists, those actually familiar with the facts, have always regarded the appeal to imperfection as a way of explaining away the absence of transitional forms with a good deal of scepticism. This was even true in the nineteenth century.

Rudwick, in *The Meaning of Fossils*, described the attitude of many of the leading paleontologists of Darwin's day.[84]

> [they] . . . were well aware of the intrinsic imperfections of the record: of the whole groups of organisms that were only preserved under exceptional circumstances (for example insects) or never preserved as fossils at all. However this did not affect their increasing confidence in the adequacy of the fossil record as evidence for the major outlines of the history of those groups which possessed readily fossilisable skeletal parts.

> John Phillips one of the leading palaeontologists in Britain . . . gave the Rede lecture at Cambridge (1860) on *Life on the Earth: Its Origin and Succession*, in which he reviewed the current evidence of palaeontology in the light of Darwin's theory. Most palaeontologists of the time would probably have agreed with him when he maintained that Darwin had grossly over-stated the case for the imperfection of the fossil record. Imperfect it certainly was, but in broad outline, and particularly for shell-bearing marine animals, it was good enough to test the plausibility of Darwin's belief in extremely slow trans-specific changes. Not only was there no positive fossil evidence for such transitions, but much more seriously it was now clear that the earliest known forms of Palaeozoic life were already highly complex organisms . . .

Darwin admitted that many of his contemporaries were not prepared to concede that the record was sufficiently imperfect to account for the gaps.[85]

> That the geological record is imperfect all will admit; but that it is imperfect to the degree required by our theory, few will be inclined to admit.

The fundamental problem in explaining the gaps in terms of an insufficient search or in terms of the imperfection of the record is their systematic character – the fact that there are fewer transitional species between the major divisions than between the minor. Between Eohippus and the modern horse (a minor division) we have dozens of transitional species, while between a primitive land mammal and a whale (a major division) we have none. And this rule applies universally throughout the living kingdom to all types of organisms, both those that are poor candidates for fossilization such as insects and those which are ideal, like molluscs. But this is the *exact reverse* of what is required by evolution. Discontinuities we might be able to explain away in terms of some sort of sampling error but their systematic character defies all explanation. If the gaps really were the result of an insufficient search, or the result of the imperfection of the record, then we should expect to find more transitional forms between mouse and whale than between dog and cat.

Moreover, there is a distinct element of tautology about the appeal to imperfection and to an insufficient search. If gradual evolution in the way conceived by Darwin did actually occur, then yes, the record is tremendously imperfect. But the question at issue is whether gradual evolution ever occurred. If the gaps cannot be adequately explained by appealing either to an insufficient search or the imperfection of the record, then this leaves a more or less saltational model of evolution as the only explanation of the gaps.

Overt saltationalism, that is proposing that new types of organisms arise suddenly, is an obvious way of avoiding the problem. Clearly, the greater the leaps that are allowed in the course of evolution, the less the requirement for transitional forms. It was precisely because Darwin was so against the idea of any sort of evolution in jumps* that he was saddled with the problem of having to explain away the mystifying absence of the "inconceivable number" of transitional forms required of any sort of gradualistic model. As Huxley pointed out the day before the publication of the *Origin*: "You have loaded yourself with an unnecessary difficulty in adopting *natura non facit saltum* so unreservedly."[86]

The tendency to view evolution in more saltational terms is inherent in the punctuational model of speciation developed recently by the American paleontologists Niles Eldredge and Stephen Jay Gould.[87] Rather than viewing the gaps as artifacts, resulting from the imperfection of the record, Eldredge and Gould believe they should be viewed as real phenomena of nature, an inevitable result of the mechanism of evolution itself. The model of evolution they propose, known as punctuated equilibrium, envisages evolution as an episodic

---

*The implausibility of evolution by macromutation, because of the enormous improbability of the sudden emergence of a new type of organism, is dealt with in Chapter Thirteen.

process occurring in fits and starts interspaced with long periods of stasis.

According to their model, new species arise rapidly in small peripherally isolated populations. During the explosive phase as the new species emerges the population undergoes rapid morphological change after which it spreads over a wide geographical area and undergoes little further change. There is a considerable amount of evidence drawn from studies of the genetics of isolated populations, such as the fruit flies of Hawaii, but many other sources as well, that this is precisely in fact how new species do arise. (See chapter four.)

If it is indeed true that new species evolve rapidly in localized geographical areas and that the populations involved are small, then obviously the chance of fossilization of transitional forms is very low. Such a model of evolution would, in Gould's own words in his book *The Panda's Thumb*, be bound to leave gaps between species.[88]

> What should the fossil record include if most evolution occurs by speciation in peripheral isolates? Species should be static through their range because our fossils are the remains of large central populations. In any local area inhabited by ancestors, a descendant species should appear suddenly by migration from the peripheral region in which it evolved. In the peripheral region itself, we might find direct evidence of speciation, but such good fortune would be rare indeed because the event occurs so rapidly in such a small population. Thus, the fossil record is a faithful rendering of what evolutionary theory predicts.

While Eldridge and Gould's model is a perfectly reasonable explanation of the gaps between species (and, in my view, correct) it is doubtful if it can be extended to explain the larger systematic gaps. The gaps which separate species: dog/fox, rat/mouse etc are utterly trivial compared with, say, that between a primitive terrestrial mammal and a whale or a primitive terrestrial reptile and an Ichthyosaur; and even these relatively major discontinuities are trivial alongside those which divide major phyla such as molluscs and arthropods. Such major discontinuities simply could not, unless we are to believe in miracles, have been crossed in geologically short periods of time through one or two transitional species occupying restricted geographical areas. Surely, such transitions must have involved long lineages including many collateral lines of hundreds or probably thousands of transitional species (see diagram on page 175). To suggest that the hundreds, thousands or possibly even millions of transitional species which must have existed in the interval between vastly dissimilar types were all unsuccessful species occupying isolated areas and having very small population numbers is verging on the incredible!

The punctuational model of Eldridge and Gould has been widely publicized but, ironically, while the theory was developed specifically to account for the absence of transitional varieties between species, its major effect seems to have been to draw widespread attention to the gaps in the fossil record. When Eldridge raised the subject with a group of science writers a few years back his views were widely reported and even reached the front page of the British newspaper *The Guardian Weekly*, but it was the absence of the transitional forms which particularly caught the attention of the reporter. According to an article entitled "Missing Believed Non-existant":[89]

> If life had evolved into its wondrous profusion of creatures little by little, Dr Eldridge argues, then one would expect to find fossils of transitional creatures which were a bit like what went before them and a bit like what came after. But no one has yet found any evidence of such transitional creatures. This oddity has been attributed to gaps in the fossil record which gradualists expected to fill when rock strata of the proper age had been found. In the last decade, however, geologists have found rock layers of all divisions of the last 500 million years and no transitional forms were contained in them.

The advent of the theory of punctuated equilibrium and the associated publicity it has generated have meant that for the first time biologists with little knowledge of paleontology have become aware of the absence of transitional forms. After this revelation of what Gould[90] has called "the trade secret of paleontology" it seems unlikely that we will see any return in the future to the old comfortable notion that the fossils provide evidence of gradual evolutionary change.

Whatever view one wishes to take of the evidence of paleontology, it does not provide convincing grounds for believing that the phenomenon of life conforms to a continuous pattern. The gaps have not been explained away.

It is possible to allude to a number of species and groups such as *Archeopteryx*, or the rhipidistian fish, which appear to be to some extent intermediate. But even if such were intermediate to some degree, there is no evidence that they are any more intermediate than groups such as the living lungfish or monotremes which, as we have seen, are not only tremendously isolated from their nearest cousins, but which have individual organ systems that are not strictly transitional at all. As evidence for the existence of natural links between the great divisions of nature, they are only convincing to someone already convinced of the reality of organic evolution.

NOTES

1. Gregory, W. K. (1951) *Evolution Emerging*, 2 vols, Macmillan Publishing Co Inc, New York, vol 2, Figure 10.5, 6(c), p 327.
2. Stanley, S. (1979) *Macroevolution*, W. H. Freeman and Co, San Francisco, p 2.
3. Marshall, N. B. (1954) *Aspects of Deep Sea Biology*, Hutchinson Publishing Co, London, p 18.
4. ibid, p 19.
5. Barnes, R. D. (1980) *Invertebrate Zoology*, 4th ed, Saunders College/Holt, Rinehart and Winston, pp 862–63.
6. Darwin, C. (1872) *The Origin of Species*, 6th ed (1962) Collier Books, New York, p 327.
7. ibid, p 330.
8. Simpson, G. G. (1960) "The History of Life" in *The Evolution of Life*, ed Sol Tax, University of Chicago Press, Chicago, pp 117–180, see p 135.
9. Morris, S. C. and Whittington, H. B. (1979) "The Animals of the Burgess Shale", *Scientific American*, 24(1): 110–120.
10. ibid, p 119.
11. Glaessner, M. F. (1961) "Pre-Cambrian Animals", *Scientific American*, 204 (3): 72–78.
12. ibid, pp 75–77.
13. Barnes, R. D. (1980) "Invertebrate Beginnings", *Paleobiology*, 6: 365–70, see p 365.
14. Darwin, C. (1881) in Darwin, F. (1888) *The Life and Letters of Charles Darwin*, 3 vols, John Murray, London, vol 3, p 248.
15. Axelrod, D. (1960) "The Evolution of Flowering Plants" in *The Evolution of Life*, op cit, p 227–306, see p 230.
16. Simpson, G. G., op cit, see p 149.
17. British Museum (Natural History) (1980) *Man's Place in Evolution*, p 20.
18. Jarvick, E. (1955) "The Oldest Tetrapods and their Forerunners", *The Scientific Monthly*, March 141–154, Figure 11, p 152.
19. Gregory, W. K. and Raven, H. C. (1942) "Studies on the Origin and Early Evolution of Paired Fins and Limbs. Part 2. A New Restoration of the Skeleton of *Eusthenopteron*", *Ann. N. Y. Acad. Sci.*, 42: 273–360.
20. Young, J. Z. (1981) *The Life of Vertebrates*, 3rd ed, Oxford University Press, Oxford, p 217.

21. Romer, A. S. (1966) *Vertebrate Paleontology*, 3rd ed, University of Chicago Press, Chicago, p71.
22. Woodward, A. S. (1898) *Vertebrate Paleontology*, Cambridge University Press, from Figure 138, p227.
23. Heilmann, G. (1926) *The Origin of Birds*, Witherby, London, from Figure 23, p37.
24. Ewer, R. S. (1965) "The Anatomy of the Theocodont Reptile *Euparkeria*", *Phil. Trans. Roy. Soc. (London)*, B248: 378–435.
25. Heilmann, op cit, pp183–191.
26. Romer, op cit, p146.
27. ibid, p166.
28. Jepson, G.L. (1970) "Bat Origins and Evolution", in *Biology of Bats*, 2 vols, ed W. Wimsatt, vol 1, pp1–64, from Figure 2, p18.
29. Gregory, op cit, from Figure 19.11, p691.
30. Young, op cit, p439.
31. Romer, op cit, p213.
32. ibid, from Figure 170, p117.
33. ibid, from Figure 172, p118.
34. Carroll, R.L. (1964) "The Earliest Reptiles", *Journal of Linnean Soc. of London (Zool.)*, 45: 61–83, from Figure 6, pp68–69.
35. Romer, op cit, p117.
36. ibid, p120.
37. Gregory, op cit, from Figure 13.11, p455.
38. Vaughn, P.P. (1955) "The Permian Reptile *Araeoscelis* Restudied", *Bull. Ms. Comp. Zool.*, 113(5): 303–467.
39. Romer, op cit, p121.
40. Kellog, R. (1936) "A Review of the Archeoceti", *Pub. Carnegie Inst. Washington*, vol 482, pp1–366.
41. Stahl, B.J. (1974) *Vertebrate History: Problems in Evolution*, McGraw-Hill Book Co, New York, p487.
42. Gregory, op cit, from Figure 20.3 (B), p730.
43. ibid, from Figure 20.40 (A), p763.
44. ibid, from Figure 20.32 (A), p756.
45. Romer, op cit, p238.
46. ibid, from Figure 367, p254.
47. Gregory, op cit, from Figure 21.25 (A), p790.
48. Romer, op cit, pp247–254.
49. Stahl, op cit, pp529–31.
50. Romer, op cit.
51. Shrock, R.R. and Twenhofel, W.H. (1953), *Principles of Invertebrate Paleontology*, McGraw-Hill Book Co, New York.
52. Darwin, op cit, p309.
53. Romer, op cit, from Figure 67, p41, and Figure 156, p108.
54. Carter, G.S. (1967) *Structure and Habit in Vertebrate Evolution*, Sidgwick and Jackson, London, Figure 1, p9.
55. Olson, L.S. and Feduccia, A. (1979) "Flight Capability and the Pectoral Girdle of *Archaeopteryx*", *Nature*, 278: 247–48.

56. Heilmann, op cit, from Figure 21, p 33.
57. Savile, D. B. O. (1957) "Adaptive Evolution in the Avian Wing", *Evolution*, 11: 212–24, see p 222.
58. Romer, op cit, pp 94–95.
59. ibid, p 95.
60. Jerison, J. H. (1973) *Evolution of the Brain and Intelligence*, Academic Press, New York.
61. Jerison (1968) "Brain Evolution and *Archaeopteryx*", *Nature*, 219: 1381–82.
62. Hardy, A. (1965) *The Living Stream*, Collins, London, Figure 59, p 201.
63. Forey, P. L. (1980) *"Latimeria*: A Paradoxical Fish", *Proc. of the Roy. Soc. of London*, B208: 369–84, see p 369.
64. Stahl, op cit, p 146.
65. Rosen, D. E., Forey, P. L., Gardiner, B. B. and Patterson, C. (1981) "Lung-fishes, Tetrapods, Palaeontology and Plesiomorphy", *Bull. Amer. Mus. Nat. Hist.*, 167: 163–275.
66. Jerison, op cit, pp 153–55 and p 213.
67. Rensch, B. (1959) *Evolution above the Species Level*, Columbia University Press, New York, p 191.
68. Storer, T. I., Usinger, R. L., Nybakken, J. W. and Stebbins, R. C. (1977) *Elements of Zoology*, 4th ed, McGraw-Hill Book Co, New York, Figure 13.9, p 212.
69. Stanley, op cit, p 39.
70. Simpson, G. G. (1953) *The Major Features of Evolution*, Columbia University Press, New York, p 263.
71. Owen, R. (1866) *Anatomy of Vertebrates*, 3 vols, Longmans, Green and Co, London, vol 3, p 795.
72. Sikes, S. K. (1971) *The Natural History of the African Elephant*, Weidenfeld and Nicolson, London, pp 2–4.
73. Newell, N. D. (1959) "The Nature of the Fossil Record", *Proc. of the Amer. Phil. Soc.*, 103 (2): 264–85, see p 267.
74. Valentine, J. W. (1977) "General Patterns of Metazoan Evolution" in *Patterns of Evolution as Illustrated by the Fossil Record*, ed A. Hallam, Elsevier Scientific Pub Co, Amsterdam, pp 27–57, see pp 34–5.
75. Cloud, P. E. (1968) "Pre-Metazoan Evolution and the Origins of the Metazoa", in *Evolution and Environment*, ed E. T. Drake, Yale University Press, New Haven, pp 1–72, see pp 25–6.
76. Darwin, op cit, p 332.
77. ibid, p 308.
78. Schopf, T. (1981) "Punctuated Equilibrium and Evolutionary Stasis", *Paleobiology* 7(2), pp 156–66, see p 160.
79. Simpson, op cit, Table 8.
80. ibid, Table 9.
81. ibid, p 143.
82. Romer, op cit, compiled from information on pp 347–96.
83. Durham, W. (1962) "The Incompleteness of our Knowledge of the Fossil Record", *Journal of Paleontology*, 41: 559–65.
84. Rudwick, M. J. S. (1972) *The Meaning of Fossils*, Neal Watson Academic Publications Inc, New York, pp 228 and 239.

85. Darwin, op cit, p 464.
86. Huxley, T. H. (1859) letter to Darwin, 23 November 1859, in Huxley, L., *Life and Letters of T. H. Huxley*, 2 vols, Macmillan and Co Ltd, vol 2, p 176.
87. Eldredge, N. and Gould, S. J. (1973) "Punctuated Equilibria: An Alternative to Phyletic Gradualism" in *Models in Paleobiology*, ed T. J. M. Schopf, Freeman, Cooper and Co, San Francisco, pp 82–115.
88. Gould, S. J. (1980) *The Panda's Thumb*, W.W. Norton and Co, New York and London, p 184.
89. *The Guardian Weekly*, 26 November 1978, vol 119, no 22, p 1.
90. Gould, op cit, p 181.

CHAPTER 9
Bridging the Gaps

It is no doubt extremely difficult even to conjecture by what gradations many structures have
been perfected . . .

Problems similiar to those involved in attempting to reconstruct
hypothetical evolutionary pathways are sometimes met with in other
fields apart from biology. Linguists, for example, are often faced
with the same sorts of problems as evolutionary biologists in working
out the relationships and origin of various language groups.[1] Just as
in biological evolution where transitional organisms which would
bridge the gaps and provide firm evidence of gradual evolutionary
descent are missing, so also in linguistics, historical and documentary
evidence of the origin and development of language groups is often
absent.

All the major Germanic languages of Europe, for example, including English, Dutch, German and Icelandic, were already well differentiated and distinct and unlinked by transitional dialects when
they first appeared in written form. Yet, despite the absence of intermediates, no linguist today doubts that all the Germanic languages
descended gradually over a period of three thousand years from an
ancestral proto-Germanic tongue. This is because they have been
able to work out in very exact detail all the semantic, syntactic and
phonetic changes which occurred along all the hypothetical pathways
through which the languages evolved. The reconstruction has been
taken to such an extent that the entire lexicon, grammar, and even the
sound of these extinct and long dead languages can be specified at
every point along all the various lineages leading back in time to the
proto-Germanic source.

These linguistic reconstructions illustrate that, even where direct
empirical evidence of a natural evolutionary relationship is missing,

Figure 9.1: Pro-avis. (from Heilman)[2]

if detailed and plausible reconstructions of the presumed evolutionary
pathway can be provided, then the assumption of an evolutionary
relationship becomes almost inescapable. As discussed in Chapter
Two, basically there are two ways by which one can justify belief in
evolution: by finding the connecting links, or by reconstructing
them. Evolution by natural selection would be established today
beyond any reasonable doubt, even without empirical evidence of

intermediates, if it had been shown that all the great divisions of nature could at least theoretically have been crossed by inventing a really convincing series of hypothetical and fully functional transitional forms. However, as we shall see, this has never been achieved.

The trouble with reconstructing hypothetical organisms is that, compared with, say, languages, organisms are tremendously complex objects and reconstructing an unknown organism, or even merely a hypothetical organ, in sufficient detail so that we could be sure it could function and survive, is a task beyond any biologist at present. Even known organisms, despite all we have learned of their physiology, biochemistry, embryology and ecology are still very much black boxes and only a fraction of their total adaptive complexity is understood. We still do not have anything approaching a complete description of even the simplest bacterial cell.

Some idea of the daunting difficulties involved in reconstructing transitional forms can be gauged by considering the vast number of factors that would have to be taken into account in the relatively simple task of redesigning a very much enlarged version of a well known organism. The German zoologist Bernhard Rensch has described the changes that would be necessary in designing a gigantic beaver-sized rat some sixty times heavier than a normal rat.[3]

> . . . a hypothetical "rat" of excessive body size, about as large as a beaver, would differ from its smaller relatives in the following characters. This animal compared with related smaller species would have a relatively smaller head, brain case (in relation to the facial bones), brain stem (in comparison to the brain as a whole), ears, and feet; a relatively shorter tail and shorter hairs; and a relatively smaller heart, liver, kidneys, pancreas, thyroid, and pituitary and adrenal glands. The weight of the bones would be relatively heavier, the facial bones relatively longer (in relation to the brain case), and the forebrain relatively larger (in relation to the brain as a whole). The retina of this giant rat would be relatively (and probably absolutely) thinner; the layer of ganglion cells and both granular layers of the eye would be less dense; the number of rods and cones would be relatively smaller. In the forebrain the cortex-7-stratificatus would be relatively larger, and the semicortex relatively smaller. The absolutely larger neurons of the brain would be less dense but would have many more dendritic ramifications. There would be equally large but definitely more numerous blood corpuscles and bone and connective tissue cells, and relatively smaller insulin-producing tissue of the pancreas. Finally,

the general metabolism (especially the rate of oxygen consumption, breathing, pulse, and blood circulation) would be decreased; the amount of blood sugar would be less, and the relative speed in locomotion would be slower. In this giant rat, the onset of maturity would be postponed, the gestation period and average length of individual age would be prolonged, and the animal would be superior in learning ability and in memory.

Any change, therefore, which on the surface may at first appear quite trivial, on closer examination would inevitably necessitate extensive reorganization of the entire anatomy and physiology of the organism. It is clear, then, that a complete reconstruction of hypothetical transitional organisms including all the details of their biology, anatomy, physiology, behaviour etc is out of the question, and it may never be possible to rigorously test any major evolutionary claim by providing fully reconstructed transitional types. Nevertheless, if the gaps were closed gradually through transitional types, as evolution implies, then at least it should be possible to provide general descriptions of the intermediates to show that they could have actually existed.

On the whole, however, even the most tentative schemes outlining a sequence of events are seldom convincing. Take, for example, the problem of the origin of birds. The flight feather of a bird is one of the most beautiful and well known of all biological adaptations.

Each feather consists of a central shaft carrying a series of barbs which are positioned at right angles to the shaft to form the vane. The barbs which make up the vane are held together by rows of barbules. From the anterior barbules, hooks project downward and these interlock with ridges on the posterior barbules. Altogether, in the flight feather of a large bird, about a million barbules cooperate to bind the barbs into an impervious vane.

The feather is a magnificent adaptation for flight. Flight feathers are remarkably light and strong and anyone who has played with one will know how easily a ruffled feather can be repaired merely by drawing it between the fingers. In addition to its lightness and strength the feather has also permitted the exploitation of a number of sophisticated aerodynamic principles in the design of the bird's wing.

One problem common to all aerofoils is turbulence, which reduces lift and causes stalling. Turbulence can be greatly cut down by the

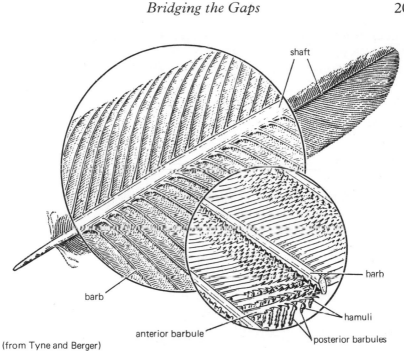

shaft

barb

barb

hamuli

anterior barbule

posterior barbules

(from Tyne and Berger)

provision of slots in the aerofoil which let through part of the air stream and tend to smooth down the flow. Aeroengineers have used this principle by placing a small subsiduary aerofoil in front of the main wing, creating the so-called Handley Page slot. The use of feathers in the design of an aerofoil lends itself admirably to the provisions of slots, and most birds' wings exploit this technique. The use of feathers also provides the bird with an aerofoil of variable geometry so that it has the ability to vary the shape and aerodynamic properties of its wing at take-off, landing, and for various different sorts of flight – flapping, gliding, soaring. In many birds, the positioning of the feathers is maintained by an intricate system of tendons which allow the feathers to twist in such a way that when the wing is raised they open like the vanes of a blind, greatly reducing resistence, but close completely on the downstroke, thus greatly improving the efficiency of flight. One need only watch the darting-backwards-and-forwards flight of the humming bird to grasp something of the excellent aerodynamic properties of the feathered aerofoil.

It is almost universally accepted by evolutionary biologists that birds evolved from reptiles, and that the feather evolved from a reptile's scale. Birds are certainly closely related to reptiles and it is difficult to see what other group of living organisms could possibly serve as hypothetical ancestors.

By what sequence of events and through what kinds of transitional states might the feather, the feathered aerofoil or wing, and avian flight have evolved? John Ostrom, an expert in this field, in a recent article in the *American Scientist* refers to the two major traditional scenarios.[5]

> Previous speculations on this question have produced two quite differ-ent scenarios. Stated very simply these are that birds began to fly "from the trees down" – or "from the ground up." The first is the widely favored and very logical "arboreal theory," . . . The second is the often ridiculed and seemingly less probable "cursorial theory," . . .

One of the classic arboreal scenarios was developed by Gerhard Heilman in his well known book *The Origin of Birds*. Heilman, as an advocate of the arboreal theory, envisaged a gliding stage preceding the development of true powered flight. The original ancestor, he suggests, was a terrestrial runner:[6]

> From being a terrestrial runner the animal now turns an arboreal climber, leaping further and further from branch to branch, from tree to tree and from the trees to the ground. Meanwhile the first toe changes to a hind toe so adapted as to grasp the branches. As the hind limbs while running on the ground have abandoned the reptilian position, they are kept closer to the body when leaping takes place, the pressure of the air acting like a stimulus, produces, chiefly on the forelimbs and the tail, a parachutal plane consisting of longish scales developing along the posterior edge of the forearms and the side edges of the flattened tail.
>
> By the friction of the air, the outer edges of the scales become frayed, the frayings gradually changing into still longer horny processes, which in course of time become more and more featherlike, until the perfect feather is produced. From wings, tail and flanks, the feather-ing spreads to the whole body. The lengthening of the penultimate phalanges of the fingers is attained by using the claws for climbing, and this elongation has been very propitious to the susequent develop-ment of the wing.
>
> The more intensive use of the arms, however, has also lengthened these, and laid claim to more powerful muscles for the movements of the same: this again has reacted on the breast bone, the two lateral halves of which have coalesced and ossified completely, forming a projecting ridge for the origin of the muscles.

Then accelerated metabolic process, finally, produced an increased caloricity protected by the feathering until the warm-blooded state was attained.

Overall, Heilman's scheme is highly speculative. He attempts no rigorous mathematical aerodynamic approach, which would give estimates of wing area, body weight, and lift at the various stages to show that his "frayed scaled" aerofoil would work and that the transition to gliding, and from gliding to powered flight was at least feasible. Indeed, Heilman's imaginary reconstruction of pro-avis (see Figure 9.1) leaves one with the distinct feeling that its wing/weight ratio would be insufficient even for gliding, let alone powered flight.

Moreover there are serious doubts about the feasibility of the transition from gliding to powered flight. As a recent article in the *American Naturalist* points out, the physical adaptations for powered flying are in opposition to those of gliding flight.[7] The aerofoil of a glider, for example, is usually a membrane attached to the body of the animal which extends out to the fore- and hindlimbs. In the case of a powered flyer, lift and thrust are usually generated by surfaces such as the wings and tail, which are some distance from the main mass of the animal.

The arboreal theory is also considered implausible by Ostrom on the grounds that all birds, including *Archaeopteryx*, exhibit various anatomical features which seem to preclude them from having descended from arboreal climbing ancestors. In his own words:[8]

> The critical point is that in order to fly, the animal first had to be able to climb. However, considering the design of modern birds, together with that of the oldest-known bird, *Archaeopteryx*, that skill may not have been part of the repertoire of primitive birds, or even of bird ancestors.

As an advocate of the "from the ground up" or "cursorial theory", Ostrom envisages bird flight to have evolved through a bipeded running and leaping stage (see Figure, page 209):[9]

> The cursorial theory postulates a sequence of stages from a primitive quadrupedal reptile, to a facultative biped, to an obligatory cursorial biped, followed by stages of elongation of the fore-limbs and enlargement of "scales" on the arms to increase their surface area, thereby forming ever larger "thrust" surfaces. Flapping action of these "proto-

wings" supposedly added thrust to that provided by the hindlimbs, resulting in greater acceleration and faster running speed. Ultimately, this is presumed to have led to flight velocities and to at least partial conversion of the forelimbs from "propellers" into wings.

There are difficulties with both the arboreal and cursorial models and the advocates of each theory see serious flaws in the alternative model. While Ostrom rejects the arboreal model on anatomical grounds, he acknowledges that the cursorial theory has not been widely accepted:[10]

> One of the key criticisms that has been leveled at this hypothesis is that, once the animal is airborne, the main thrust source (ie traction of the hind feet against the ground) would be lost and velocity would diminish.

An even more serious difficulty would be:[11]

> the miniscule amount of additional thrust that could have been generated by those earliest enlarged "scales" on the incipient "wing" pushing against the air. Certainly this could not have been anywhere near enough additional thrust to produce a measurable increase in running speed, and thus be selected for.

As he admits:[12]

> The cursorial theory of bird flight origins has received virtually no acceptance, apparently for several very good reasons . . . including the seemingly impossible "bootstrap" effort required for the animal to lift itself by means of flapping proto-wings.

Recently, the cursorial theory has received renewed support from aerodynamic studies[13] carried out by a research team at Northern Arizona University. Rejecting the traditional approach, adopted by most advocates of the cursorial theory, which has always stressed selection for increasing thrust and lift as the primary determinants in the evolution of the avian wing, the Arizona group adopted an alternative premise, that it was selection for control of body orientation that was the most critical factor in the evolution of avian flight. In their view, wings evolved initially as balancing organs to give stability while running and to assist in the control of body position as the animal jumped after its prey. Like Ostrom, the Arizona group envisages pro-avis to have been a fast running bipedal insectivore, but one which leapt to catch flying prey in its mouth rather than trapping them in an insect net.

Using a fifteen by three centimeter one hundred gram cylinder they calculated that even a very slight degree of lift consequent on only minor extensions of the forelimb and tail would greatly increase the ability of a running bipedal animal to control its orientation while running and particularly during leaping after prey, thus enhancing its foraging efficiency.

They speculate that once the animal had achieved proto-wings, which enhanced lift and aerodynamic control, then the advantage gained from greater foraging ability would ensure rapid selection for even greater lift and quickly lead to the acquisition of powered flight. They also point out that the use of the limbs in controlling pitch and roll during running or jumping would resemble the lazy figure-of-eight characteristic of a typical avian power stroke.

Although plausible to some degree, like other models it raises a number of problems. An obvious difficulty is that no known animal regularly catches flying insects by leaping after them in the way envisaged in this model. Nearly all insectivorous vertebrate species take their prey on the ground. Only the most skilled flyers, the bats and a few species of bird, are able to capture insects in the air. One insectivorous bird curiously reminiscent of the cursorial pro-avis is the roadrunner of Mexico and the southwestern states of the United States. This interesting bird can run at twenty-six miles an hour but flies poorly and seldom takes to the wing; unlike pro-avis it never attempts to leap after flying insects. One suspects that catching flying insects is more difficult than one might have imagined, and the ecological nitch envisaged for pro-avis in the Arizona group's model is not particularly attractive.

Although many variants of both the arboreal and cursorial theories have been proposed over the past century, to date no overall scheme has ever been developed which has not seemed implausible to some degree to a significant number of authorities. Moreover, on top of the still unsolved general question as to whether birds flew up or down there are a host of more specific problems, such as, for example, the difficulty of explaining the origin of the feather.

The central difficulty with all gradual schemes for the evolution of the feather is that any aerofoil constructed out of feathers will only work if the feathers are strong, capable of resisting deformation and capable of forming an impervious vane. Moreover, there has to be a sufficient number of feathers to provide a sufficient surface area to achieve the requisite degree of lift. An aerofoil based on the feathered design has, therefore, to satisfy a number of quite stringent criteria before it can function and create lift. It is significant in this regard that every single flying bird, from Archeopteryx on, has possessed a highly developed aerofoil consisting of a complex arrangement of fully developed flight feathers.

According to Heilman the original impervious vane which supported these pre-avian species as they glided was a set of "longish scales developing along the posterior edge of the forearms and the side edges of the flattened tail". Then, he continues:[14]

> By the friction of the air the outer edges became frayed, the fraying gradually changing into still longer horny processes which in the course of time became more and more feather like.

It is at this point, when the actual evolution of the feather is envisaged, that Heilman's scheme begins to look particularly implausible, for it is very difficult to understand what the adaptive value of frayed scales would be to a gliding organism when any degree of fraying would make the scales pervious to air, thereby decreasing their surface area and lift capacity. All known organisms which have adaptations for gliding among fish, frogs, reptiles, and mammals present a continuous unbroken surface to the air. It would seem reasonable to believe that selection for gliding in a hypothetical pro-avis would always tend to increase the impervious surface area of its wing and decrease the tendency to fray.

While Heilman envisages the evolution of feathers as being directly involved in the acquisition of flight, Ostrom has speculated that perhaps they first arose as a device to catch insects:[15]

> Is it possible that the initial (pre-*Archaeopteryx*) enlargement of feathers on its hand might have been to increase the hand surface area, thereby making it more effective in catching insects? Continued selection for larger feather size could have converted the entire forelimb into a large, light-weight "insect net." It is not difficult to visualize how advantageous these paired "insect nets" would be in snaring leaping insects, or even in batting down escaping flying insects.

It seems particularly doubtful that "feathers" evolving to form an insect net would provide the basis for an impervious aerofoil. A net must be (as anyone who has tried to swat a mosquito will have discovered) pervious to the air. If a reptile's scale ever did evolve in this direction (and no other living organism has ever possessed such a remarkable structure), it would surely be pervious to air and unsuitable for any sort of flight.

It is not easy to see how an impervious reptiles scale could be converted gradually into an impervious feather without passing through a frayed scale intermediate which would be weak, easily deformed and still quite permeable to air. It is true that basically a feather is indeed a frayed scale – a mass of keratin filaments – but the filaments are not a random tangle but are ordered in an amazingly complex way to achieve the tightly intertwined structure of the feather. Take away the exquisite coadaptation of the components, take away the co-adaptation of the hooks and barbules, take away the precisely parallel arrangement of the barbs on the shaft and all that is left is a soft pliable structure utterly unsuitable to form the basis of a stiff impervious aerofoil. The stiff impervious property of the feather which makes it so beautiful an adaptation for flight, depends basically on such a highly involved and unique system of coadapted components that it seems impossible that any transitional feather-like structure could possess even to a slight degree the crucial properties. In the words of Barbara Stahl, in *Vetebrate History: Problems in Evolution*, as far as feathers are concerned[17] "how they arose initially, presumably from reptiles scales, defies analysis."

The evolution of birds is far more complex than the above discussion implies. In addition to the problem of the origin of the feather and flight, birds possess other unique adaptations which also seem to defy plausible evolutionary explanations. One such adaptation is the avian lung and respiratory system.

In all other vertebrates the air is drawn into the lungs through a system of branching tubes which finally terminate in tiny air sacs, or alveoli, so that during respiration the air is moved in and out through the same passage.

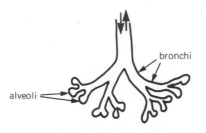

In the case of birds, however, the major bronchi break down into tiny tubes which permeate the lung tissue (see Figure 9.2). These so-called parabronchi eventually join up together again, forming a true circulatory system so that air flows in one direction through the lungs.

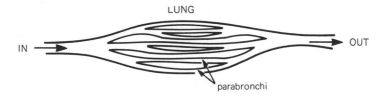

This unidirectional flow of air is maintained during both inspiration and expiration by a complex system of interconnected air sacs in the bird's body which expand and contract in such a way so as to ensure a continuous delivery of air through the parabronchi. The existence of this air sac system in turn has necessitated a highly specialized and unique division of the body cavity of the bird into several compressible compartments. Although air sacs occur in certain reptilian groups, the structure of the lung in birds and the overall functioning of the respiratory system is quite unique. No lung in any other vertebrate species is known which in any way approaches the avian system. Moreover, it is identical in all essential details in birds as diverse as humming birds, ostriches and hawks.

Just how such an utterly different respiratory system could have

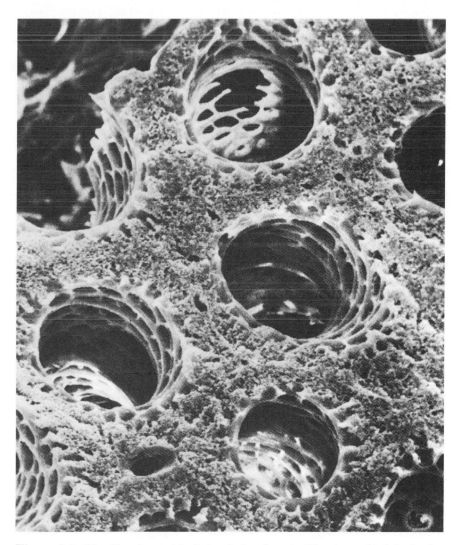

Figure 9.2: The Parabronchi of the Avian Lung. *The tiny cylindrical tubes which permit the unidirectional flow of air through the lungs. Each tube is about .5mm in diameter.* (from Düncker)[18]

evolved gradually from the standard vertebrate design is fantastically difficult to envisage, especially bearing in mind that the maintenance of respiratory function is absolutely vital to the life of an organism to the extent that the slightest malfunction leads to death within minutes. Just as the feather cannot function as an organ of flight until the hooks and barbules are coadapted to fit together perfectly, so the avian lung cannot function as an organ of respiration until the parabronchi system which permeates it and the air sac system which guarantees the parabronchi their air supply are both highly developed and able to function together in a perfectly integrated manner.

Moreover, the unique function and form of the avian lung necessitates a number of additional unique adaptations during avian development. As Dunker,[19] one of the world's leading authorities in this field, explains, because, first, the avian lung is fixed rigidly to the body wall and cannot therefore expand in volume and, second, because of the small diameter of the lung capillaries and the resulting high surface tension of any liquid within them, the avian lung cannot be inflated out of a collapsed state as happens in all other vertebrates after birth. In birds, aeration of the lungs must occur gradually and starts three to four days before hatching with a filling of the main bronchi, air sacs and parabronchi with air. Only after the main air ducts are already filled with air does the final development of the lung, and particularly the growth of the air capillary network, take place. The air capillaries are never collapsed as are the alveoli of other vertebrate species; rather, as they grow into the lung tissue, the parabronchi are from the beginning open tubes filled either with air or fluid (which is later absorbed into the blood capillaries).

In attempting to explain how such an intricate and highly specialized system of correlated adaptations could have been achieved gradually through perfectly functional intermediates, one is faced with the problem of the feather magnified a thousand times. The suspicion inevitably arises that perhaps no functional intermediate exists between the dead-end and continuous through-put types of lung. The fact that the design of the avian respiratory system is essentially invariant in ALL birds merely increases one's suspicion that no fundamental variation of the system is compatible with the preservation of respiratory function. One is irresistibly reminded of Cuvier's view that the great divisions of nature are grounded in necessity and that intermediates cannot exist because such forms are incoherent and nonfunctional. In his own words:[20]

Nature . . . has been settled in . . . all those combinations which are not incoherent and it is these incompatibilities, this impossibility of the coexistence of one modification with another which establish between the diverse groups of organisms those separations, those gaps, which mark their necessary limits . . .

The avian lung and the feather bring us very close to answering Darwin's challenge:[21]

If it could be demonstrated that any complex organ existed which could not possibly have been formed by numerous, successive, slight modifications, my theory would absolutely break down.

In addition to the feather and the avian lung there are many other unique features in the biology of the birds, in the design of the heart and cardiovascular system, in the gastrointestinal system and in the possession of a variety of other relatively minor adaptations such as, for example, the unique sound producing organ, the syrinx, which similarly defy plausible explanation in gradualistic terms. Altogether it adds up to an enormous conceptual difficulty in envisaging how a reptile could have been gradually converted into a bird.

What we seem to have, then, is a very interesting coincidence – a great empirical discontinuity in nature between reptiles and birds which seems to coincide with a major conceptual discontinuity in our ability to conceive of functional intermediates through which the gap might have been closed.

The difficulty of envisaging how evolutionary gaps were closed does not stop with birds: Take the case of the bats. The first known bat which appeared in the fossil record some sixty million years ago had as completely developed wings as modern forms. As in the case of birds, how could the development of the bats' wings and capacity for powered flight have come about gradually? Darwin in the *Origin* suggests, like Heilman, that bat wings evolved gradually from an original gliding device. It is worth quoting Darwin's reasoning at length, for it again illustrates that explanations in this area often leave a lot to be desired:[22]

Look at the family of squirrels; here we have the finest gradation from animals with their tails only slightly flattened, and from others, as Sir Richardson has remarked, with the posterior part of their bodies rather wide and with the skin on their flanks rather full, to the so-called flying squirrels; and flying squirrels have their limbs and even the base of the tail united by a broad expanse of skin, which serves as a parachute and allows them to glide through the air to an astonishing distance from tree to tree. We cannot doubt that each structure is of use to each kind of squirrel in its own country, by enabling it to escape birds or beasts of prey, to collect food more quickly, or, as there is reason to believe, to lessen the danger from occasional falls. But it does not follow from this fact that the structure of each squirrel is the best that it is possible to conceive under all possible conditions. Let the climate and vegetation change, let other competing rodents or new beasts of prey immigrate, or old ones become modified, and all analogy would lead us to believe that some at least of the squirrels would decrease in numbers or become exterminated, unless they also became modified and improved in structure in a corresponding manner. Therefore, I can see no difficulty, more especially under changing conditions of life, in the continued preservation of individuals with fuller and fuller flank-membranes, each modification being useful, each being propagated, until by the accumulated effects of this process of natural selection, a perfect so-called flying squirrel was produced.

Now look at the *Galeopithecus* or so-called flying lemur, which formerly was ranked amongst bats, but is now believed to belong to the Insectivora. An extremely wide flank-membrane stretches from the corners of the jaw to the tail, and includes the limbs with the elongated fingers. This flank-membrane is furnished with an extensor muscle. Although no graduated links of structure, fitted for gliding through the air, now connect the *Galeopithecus* with the other Insectivora, yet there is no difficulty in supposing that such links formerly existed, and that each was developed in the same manner as with the less perfectly gliding squirrels; each grade of structure having been useful to its possessor. Nor can I see any insuperable difficulty in further believing that the membrane connected fingers and fore-arm of the *Galeopithecus* might have been greatly lengthened by natural selection; and this, as far as the organs of flight are concerned, would have converted the animal into a bat.

In the above quote Darwin takes us up to a highly developed gliding form, *Galeopithecus*. But this animal is certainly no bat and, although admittedly its fingers are somewhat elongated, they form a normal functional hand. From *Galeopithecus* to bat there is a massive jump and, although Darwin tells us "I see no insuperable difficulty", he does not explain how the transition was made. Thus he avoids the essential problem of bat evolution, envisaging functional intermediate stages between a normal forelimb and a wing. The forelimbs and hand of *Galeopithecus* are basically the same as in any normal quadruped, while in the bat the fingers are greatly extended and consequently no longer serviceable as a normal hand.

Jepsen recently commented on the supposed intermediate status of *Galeopithecus*:[23]

> Although the gliding dermopteran (*Galeopithecus*) which is wholly misnamed 'the flying lemur,' is said to be ". . . almost a bat in some respects" (Allen, 1939) or illustrative of ". . . an intermediate stage in the development of flight" (Romer, 1959) it may be as closely related to rats as to bats. It is almost as logical to think of the flying lemurs as being derived from bats as it is to entertain the idea of a *Galeopithecus*-bat lineage.

He also takes the view that a gliding stage, such as might be represented by *Galeopithecus*, could not possibly lead to true flight:[24]

> Morphologically and genetically and phylogenetically the distance from a gliding habit to a bat-flying habit among known mammals is so immense that a development of the former may almost be said to preclude the probability of further development in the same phyletic line to the latter.

We seem to be forced, in trying to envisage bat evolution, to imagine a succession of small mammalian species in which the fingers gradually lengthened, resulting in loss of normal forelimb function, before the necessary development of wings and specialized muscle to sustain powered flight had been attained.

Would a primitive wing, far less efficient than a modern bat's, allowing only very restricted movement through the air be of such selective advantage that an organism would sacrifice its forelimbs in its favour? If the wings of bats really did evolve gradually from gliding organs then presumably at some stage during their evolution we are forced to accept that such a choice must have been made. As one authority on bat evolution, James Smith, recently commented at a symposium on vertebrate evolution:[26]

> Certainly early stages in the development of the wing would have allowed both volant and relatively normal quadrupedal locomotion. However, the continued development of the wing, in this manner, eventually would have produced an ungainly and clumsy structure.

Again, as in the case of pro-avis, doubts arise. Would such a transitional form really be able to function? Its hands would be comparatively useless and yet its half-developed wings could hardly have supported anything more than rudimentary flight. One's suspicions of the feasibility of such a transitional form are not diminished by the fact that the diversity of gliding forms is very great, yet no known type approaches the condition of the bat, which suggests that the ecological niche available to small mammals half way to true flight is not very attractive. As Jepsen concedes:[28]

> No one has successfully proposed any kind of selection pressure that would be effective in the change from one niche to the other; whether the bridging group would be pulled by advantages in the new milieu or pushed by disadvantages in the old.

The large marine aquatic vertebrates such as the icthyosaurs (the fish reptiles), the seals and the whales, etc appear suddenly in the fossil record already fully differentiated, and once again only the most general explanations of how such transitions occurred are offered. Detailed analyses providing relatively complete blueprints of the transitional stages, including descriptions of skeletal and muscular morphology as well as cardiovascular and reproductive physiology and behaviour, to show clearly that their gradual evolution could have occurred are never provided. Instead what we get is often like the following, taken from a quite old text but still typical:[29]

> We may begin with an animal like the stoat that occasionally jumps into the water and swims well. The next step may be illustrated by the otter, that is thoroughly at home in the river and may swim for miles out to sea, yet remains equally at home on land. On the next level may be placed the almost exterminated sea-otter (*Enhydris*) of the North Pacific, whose hind feet are suited only for swimming. Then we reach the progressive series represented by the sea-lion, walrus and seals – the last named being almost as thoroughly aquatic as the whales, except that they bring forth their young on the shore and nurse them there.

Such a scheme has a certain element of plausibility but, like Heilman's description of the evolution of the feather, it tends to slur over crucial difficulties and is far too vague to be subjected to serious criticism. The impression conveyed that there is a continuity of types between stoat and whale is utterly misleading. Between sea otter and seal and between seal and whale there are enormous discontinuities unbridged by any known or extinct form. As soon as one attempts a detailed reconstruction of the transitions one finds oneself again in the unsatisfying realm of pro-avis. D. Dewar, a leading anti-envolutionist in the 1930s, challenged his zoological colleagues to provide detailed blueprints of intermediate forms:[30]

But I do not challenge evolutionists to make sketches of actual ancestors of the whales. I ask for drawings of skeletons of possible intermediates. So far no one has taken up my challenge. Mr Arnold Lunn is more fortunate. It is recorded ("Science and the Supernatural" by Lunn and Haldane, p320) that in response to this challenge Professor J. B. S. Haldane pleaded that he is not a good sketcher, but that his drawings would be rather like caricatures of dugongs and seals. Now the dugong, being as well adapted to swimming with the tail as the whale, can scarcely be called an intermediary between the latter and a land mammal. The seal, although adapted to existence both on land and in water, is not anatomically intermediate between a whale and a land animal. Let us notice what would be involved in the conversion of a land quadruped into, first a seal-like creature and then into a whale. The land animal would, while on land, have to cease using its hind legs for locomotion and to keep them permanently stretched out backwards on either side of the tail and to drag itself about by using its fore-legs. During its excursions in the water, it must have retained the hind legs in their rigid position and swim by moving them and the tail from side to side. As a result of this act of self-denial we must assume that the hind legs eventually became pinned to the tail by the growth of membrane. Thus the hind part of the body would have become like that of a seal. Having reached this stage, the creature, in anticipation of a time when it will give birth to its young under water, gradually develops apparatus by means of which the milk is forced into the mouth of the young one, and meanwhile a cap has to be formed round the nipple into which the snout of the young one fits tightly, the epiglottis and laryngeal cartilage become prolonged downwards so as tightly to embrace this tube, in order that the adult will be able to breath while taking water into the mouth and the young while taking in milk. These changes must be effected completely before the calf can be born under water. Be it noted that there is no stage intermediate between being born and suckled under water and being born and suckled in the air. At the same time various other anatomical changes have to take place, the most important of which is the complete transformation of the tail region. The hind part of the body must have begun to twist on the fore part, and this twisting must have continued until the sideways movement of the tail developed into an up-and-down movement. While this twisting went on the hind limbs and pelvis must have diminished in size, until the latter ceased to exist as external limbs in all, and completely disappeared in most, whales.

Every textbook of evolution asserts that reptiles evolved from amphibia but none explains how the major distinguishing adaptation of the reptiles, the amniotic egg, came about gradually as a result of a successive accumulation of small changes. The amniotic egg of the reptile is vastly more complex and utterly different to that of an amphibian. There are hardly two eggs in the whole animal kingdom which differ more fundamentally:

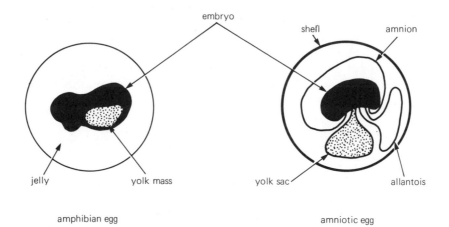

amphibian egg   amniotic egg

The diagram above illustrates some of the main distinguishing features of the amniotic egg: the tough impervious shell, the two membranes, the amnion which encloses a small sac in which the embryo floats, and the allantois in which the waste products formed during the development of the embryo accumulate, and the yolk sac containing the food reserve in the form of the protein albumen. None of these features are found in the egg of any amphibian.

The evolution of the amniotic egg is baffling. It was this decisive innovation which permitted for the first time genuinely terrestrial vertebrate life, freeing it from the necessity of embryological development in an aquatic environment. Altogether at least eight quite different innovations were combined to make the amniotic revolution possible: the formation of a tough impervious shell; the formation of the gellatinous egg white (albumen) and the secretion of a special acid to yield its water; the excretion of nitrogenous waste in the form of water insoluble uric acid; the formation of the amniotic cavity in which the embryo floats (This is surrounded by the amniotic membrane which is formed by an outgrowth of mesodermal tissue. Neither the amniotic cavity nor the membrane which surrounds it has any homologue in any amphibian); the formation of the allantois from the future floor of the hind gut as a container for waste products and later to serve the function of a respiratory organ; the development of a tooth or caruncle which the developed embryo can utilize to break out of the egg; a quantity of yolk sufficient for the needs of the embryo till hatching; changes in the urogenital system of the female permitting fertilization of the egg before the hardening of the shell.

The problem of the origin of the amniotic system is even more enigmatic considering that the basic problem it solves, in freeing reproduction from dependency on a pool of water, has been solved in the amphibia by much less radical means, by merely exploiting the basic amphibian egg. Some amphibian eggs have a tough gelatinous skin which will stand a certain degree of desiccation, others are live bearing. Certain amphibia are therefore quite independent of water for reproduction.

The origin of the amniotic egg and the amphibian – reptile transition is just another of the major vertebrate divisions for which clearly worked out evolutionary schemes have never been provided. Trying to work out, for example, how the heart and aortic arches of an amphibian could have been gradually converted to the reptilian and mammalian condition raises absolutely horrendous problems.

The living world is full of innumerable other systems, particulary among the insects and invertebrates, for which gradual evolutionary explanations have never been provided. A particularly fascinating case is the mating flight of the dragonfly. The male flies ahead of the female and grips her head with terminal claspers. The female then bends her abdomen forward and receives the sperm from a special copulatory organ which is situated toward the front on the undersurface of the abdomen of the male dragonfly and which he fills with semen from the true reproductive aperture before the start of the mating flight. This strange manoeuvre, which seems a curiously roundabout way to bring sperm to egg, depends on the unique and complex machinery which forms the male copulatory organ. Although in its detailed structure it varies enormously in different species, the fundamental design of this extraordinary complex organ is essentially the same in all species of dragonfly. No other insect possesses anything remotely like it, nor is it led up to gradually by a sequence of simpler transitional structures.

As Tillyard remarked:[31]

> The copulatory apparatus of the male Dragonfly is one of the most remarkable structures in the Animal Kingdom. The "palpal organ" on the pedipalp of the male Spider, and the hectocotylous arm of the Cephalopod Mollusc, extraordinary as they are, do not defy all explanation, since in each case they are modifications of an appendage already present. But the apparatus of the male Dragonfly is not homologous with any known organ in the Animal Kingdom; it is not derived from any pre-existing organ; and its origin, therefore, is as complete a mystery as it well could be.

An interesting example of a very widespread invertebrate phenomenon, the origin of which is in most cases difficult to account for in gradualistic evolutionary terms, is that of metamorphosis. Many invertebrates undergo a dramatic metamorphosis between the egg and adult form. As described in Chapter Seven, in the case of certain types of insect such as butterflies, beetles, bees and ants, which undergo what is termed complete metamorphosis during a quiescent pupation stage, the transformation involves virtually the complete dissolution of all the organ systems of the larva and their reconstitution *de novo* from small masses of undifferentiated embryonic cells called the imaginal discs. In other words, one type of fully functional organism is broken down into what amounts to a nutrient broth from which an utterly different type of organism emerges.

The insects are by no means unique. The crustacean *Sacculina*, a parasite of the edible crab, has a life history which involves a remarkable metamorphosis. The egg hatches into a typical free swimming crustacean larva, which then develops a bivalve shell and comes to resemble a small water flea. During this stage the larvae develop an organ for piercing the integument of a crab. On entering the crab it undergoes one of the most extraordinary transformations in nature. From being a crustacean-like organism it gradually changes, losing all its internal structure and organs, into an amorphous mass of cells which sends out root-like processes into the tissue of the crab. These processes, which resemble fungal fibres, ramify through the crab tissue absorbing nutrients and convey them back to the main mass of the organism which at this stage is little more than an egg producing bag.

The life history of some parasites, which are in themselves astonishing enough, often involve what amounts to a number of metamorphoses. Consider the life cycle of the liver fluke. The adult lives in the intestine of a sheep. After the eggs are laid they pass with the faeces onto the ground. The eggs hatch, giving rise to small ciliated larvae which can swim about in water. If the larvae are lucky they find a pond snail: they must do this to survive, for the snail is the vehicle for the next stage in the life cycle of the liver fluke. Having found a snail the larvae finds its way into the pulmonary chamber or lung. Here it loses its cilia and its size increases. At this stage it is known as a sporocyst. While in this condition it buds off germinal cells into its body cavity which develop into a second type of larvae known as rediae. These are oval in shape, possessing a mouth and stomach and a pair of protuberances which they use to move about. The rediae eventually leave the sporocyst, entering the tissue of the snail, after which they develop into yet another larval form known as cercariae which appear superficially to resemble a tadpole. Using their long tails these tadpole-like larvae work their way through and eventually out of the snail and onto blades of grass, where each larva sheds its tail and encases itself in a sheath. Eventually they are eaten by a sheep. Inside the sheep they find their way to the liver where they develop sexual organs and mature into the adult state. They finally leave the sheep's liver and migrate to the intestine where they mate and so complete their extraordinary life cycle.

In the case of many of the more dramatic invertebrate metamorphoses not even the vaguest attempts have been made to provide hypothetical scenarios explaining how such an astonishing sequence of transformations could have come about gradually as a result of a succession of small beneficial mutations. As leading parasitologist Asa Chandler admitted:[32]

> It would be difficult, if not impossible, to explain, step by step, the details of the process of evolution by which some of the highly specialized parasites reached their present condition.

The life cycles of the liver fluke or *Sacculina* and the metamorphoses of insects are merely representatives of a vast number of complex phenomena which have never been adequately accounted for in terms of a slow accumulation of beneficial mutations. As long standing critic of Darwinian orthodoxy, Ludwig Bertalanffy, confessed in 1969 at the symposium 'Beyond Reductionism':[33]

> I, for one, in spite of all the benefits drawn from genetics and the mathematical theory of selection, am still at a loss to understand why it is of selective advantage for the eels of Comacchio to travel perilously to the Sargasso sea, or why *Ascaris* has to migrate all around the host's body instead of comfortably settling in the intestine where it belongs; or what was the survival value of a multiple stomach for a cow when a horse, also vegetarian and of comparable size, does very well with a simple stomach or why certain insects had to develop those admirable mimicries and protective colorations when the common cabbage butterfly is far more abundant with its conspicuous white wings. One cannot reject these and innumerable similar questions as incompetent; if the selectionist explanation works well in some cases, a selectionist explanation cannot be refused in others.

The origin of a good number of behaviour patterns, especially among insects, involving a repertoire of several complex steps, each one being crucial to the success of the whole ritual, also defy really plausible evolutionary explanations. The works of the French entomologist Henri Fabre are full of examples such as the digger wasps, mason bees, dung beetles and the mating semaphore dance of certain spiders.

Some sorts of behaviour patterns even seem to involve what appears to be a form of interspecies altruism. Anyone who has watched the spider-hunting wasps at work is forced to ask with zoologist Garrett Hardin[34] "why under Darwinian principles doesn't the spider try to escape its nemesis the wasp?" Naturalists since the time of Henri Fabre have speculated over the same point. William S. Bristow describes the wasp *Pompilus plumbeus* hunting the spider *Arctosa perita*:[35]

> An *Arctosa* was put in a tube of diameter similar to that of her burrow. A *Pompilus* was then transferred to the same tube. Surely the spider would leap at the wasp and destroy it? No, at the first touch of the vibrating antennae a forward lunge by the spider was checked and she stayed still with her legs crossed and entwined round her cephalothorax in a completely unnatural pose whilst the wasp curled its abdomen

round to inflict a sting beneath her in the region of her sternum, thus a paralysis caused by what I can only describe as fear was replaced by a paralysis caused by poison.

Petrunkevitch, a world authority on spiders, describes the giant wasp *Pepsis marginata* hunting the tarantula *Cyrtopholis portoricae* in similar terms:[36]

> It is a classic example of what looks like intelligence pitted against instinct, the victim although fully able to defend itself, submits unwittingly to its destruction.

What possible survival value, one wonders, could accrue to the spider by such curiously altruistic behaviour? And it is not just one or two species of spiders that fall so easily to the wasps; practically every group of spiders is preyed on by a particular species of spider hunting wasp.

The work of Chrystal[37] demonstrates that the larva of the wood wasp *Sirex* is also peculiarly accommodating towards its predator, the parasitic wasp *Ibalia*. *Sirex* bores a hole in the trunk of a conifer, in which it deposits its egg. The egg yields a grub which feeds on the wood. As the grub feeds on the wood it gradually bores a tunnel. After some years the grub turns into a pupa which finally yields the adult wasp, which, using its powerful jaws, bites its way out of the tree. The *Ibalia* using the hole bored by the *Sirex* lays its egg in the *Sirex* grub. The *Ibalia* grub gradually consumes the tissues of the *Sirex* grub but does not eat the vital organs until last, thus ensuring a fresh supply of meat until its development, which takes three years, is complete. The presence of the *Ibalia* changes the behaviour of the *Sirex*. Normally the *Sirex* larva bores deeply into the wood but when infected by the *Ibalia* it bores towards the surface. This is a vital behavioural change for *Ibalia* because it has comparatively weak jaws and would be unable to bore as far through the wood as *Sirex* to escape from the trunk. Yet another example of interspecific altruism? What conceivable value can the *Sirex* grub gain by changing the direction of its boring? By what curious sequence of small evolutionary steps did the *Ibalias'* predatory habit induce this vital behavioural change?

Even bacteria provide examples of complex systems which pose a challenge to gradualistic explanations. Take, for example, the bacterial flagellum. This tiny microscopic hair, which has been observed by light microscopy for more than one century, has only very recently been elucidated. As a result, we now know that it has a completely different molecular structure to the cilium (described in Chapter Five) and recent research into the structure and function of this fascinating organelle has revealed that it possesses a remarkable property. It is the only structure in the entire living kingdom which exhibits a true rotary motion. Howard Berg described some of the latest research on the bacterial flagellum in an excellent *Scientific American* article in 1975.

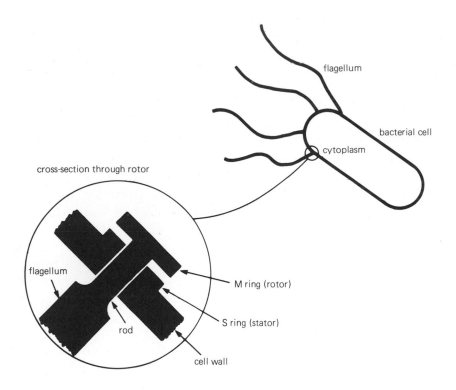

Figure 9.3: The Rotary Motor of the Bacterial Flagellum. *This is the only rotary device known in nature. According to Howard Berg the torque is generated by a translocation of ions through the M ring (a disc mounted rigidly on the rod of the flagellum and free to rotate in the cytoplasm), where they interact with charges on the surface of the S ring (which is mounted on the cell wall), imparting a rotary motion to the flagellum.* (from Berg)[38]

Unlike cilia which beat by the propagation of a wave from their base to their tip, the helical filaments which comprise the bacterial flagellum rotate rapidly like propellers and are driven by a reversible motor at their base. In Berg's words:[39]

> The evidence at hand suggests a model for the rotary motor in which the torque is generated between two elements in the basal body, the M ring and the S ring [see Figure 9.3]. The rod (which is connected to the filament by the hook) is fixed rigidly to the M ring, which rotates freely in the cytoplasmic membrane. The S ring is mounted on the cell wall. (Note that the motor must be mounted rigidly somewhere on the cell wall if the torque is to be applied.) The torque could be generated by the active translocation of ions through the M ring to interact with charged groups on the surface of the S ring.

The bacterial flagellum and the rotory motor which drives it are not led up to gradually through a series of intermediate structures and, as is so often the case, it is very hard to envisage a hypothetical evolutionary sequence of simpler rotors through which it might have evolved gradually.

Botany offers many examples of complex adaptations which have never been explained convincingly in gradualistic terms. A classic example is the pollination mechanism of the orchid *Coryanthes* described by Darwin as being:[40]

> effected in a manner that might perhaps have been inferred from their structure, but would have appeared utterly incredible had it not been repeatedly witnessed by a careful observer.

One part of this remarkable flower consists of a special bucket which is filled with a watery fluid secreted by special glands situated just above the rim of the bucket. Another gland situated on a part of the flower directly above the bucket secretes a fluid which is irresistible to bees; in their jostling for the secretion some inevitably tumble down into the fluid in the bucket below. When the bucket is full it overflows through a specially constructed spout which is also the only means of escape for any bee which happens to fall into the bucket. The spout is roofed over by a special lid which bears the stigma (the female pollen receiving organ) and a number of pollen masses. The lid fits tightly over the spout forming a narrow passage so that any bee attempting to escape from the bucket must exert considerable force to squeeze its way out through the spout, and in the process must inevitably brush first against the stigma and then against the viscid discs of pollen.[41] Thus the bee carries with it from the flower fine masses of pollen which inevitably, as it repeats the whole curious ritual, will be deposited on the stigma of the same or an adjacent flower, thereby ensuring that pollination occurs.

The adaptations by which certain insectivorous plants, such as the venus fly trap or the pitcher plant, first lure, then trap and digest their insect prey are perhaps even more incredible. Lloyd, who reviewed the extensive literature on carnivorous plants, commented:[42]

> About the origin and evolution of the carnivorous plants, however, much as these questions may intrigue the mind, little can be said, nor have I attempted to discuss them. How the highly specialized organs of capture could have evolved seems to defy our present knowledge.

As Wardlaw confesses:[43]

> Special adaptive features such as those exemplified by the plants of special habitats, climbing plants, insectivorous plants, the numerous cunning floral arrangements that ensure cross-pollination, and so on virtually ad lib., seem to the writer to be difficult to account for adequately in terms of a sequence of small random variations, and natural selection . . .

and as in so many other instances:[44]

> When one tries to account for the ontogenesis and phylogenesis of some special adaptive feature, one has no conviction that any adequate detailed explanation has yet been advanced.

It is common for evolutionary biologists to deprecate the tendency to discuss the inadequacies of natural selection with reference to special cases. As Julian Huxley complained:[45]

> It is perhaps unfortunate that the study of adaptations has been so closely associated with highly specialized and striking cases of the "wonders of nature" type, such as the almost fantastic contrivances of certain orchids which secure insect-pollination.

Yet, as Wardlaw reminds us:[46]

It is an inescapable fact that there are indeed very large numbers of these special cases both in the Plant and Animal Kingdoms which are not satisfactorily accommodated in the omnibus of evolutionary doctrine.

Indeed, practically every group of organisms possess complex adaptations equivalent in many ways to the feather and the pollination system of the *Coryanthes* which are not led up to gradually through a continuum of functional intermediate structures. The click beetles (the Elateridae) have a unique jack knife jumping device; the fleas, a complex reproductive system, more complex even than that of the dragonfly, unparalleled in any other insect order; the Echinoderms (starfishes and sea urchins), an incredibly complex water vascular system; bacteria have their tiny rotary motor; and so on.

But even adaptations of a far less spectacular kind often present a challenge to gradualistic explanations. For example, how the mechanical seed dispersal mechanisms of common garden plants like the Touch-me-not or Geranium might have gradually evolved through a succession of tiny advantageous steps is surprisingly problematical.

It is only when adaptations become very simple that envisaging how they may have come about becomes relatively straightforward. It is easy, for example, to imagine the evolution by natural selection of a new strain of bacteria resistant to penicillin or the evolution of the white coloration of a polar bear. Even quite complex behaviour patterns, such as the sophisticated stratagems adopted by the males of many species in ritual combat or the alarm call of birds which risks the life of the individual for the survival of the flock, can be reduced to selectionist explanations. But in all such cases there is an obvious functional continuum, along which selection can move step by step to achieve a particular adaptive goal. Indeed, in many cases, given the ubiquity of selection pressures which will automatically ensure that every species attempts to maximize its survival potential, and given a functional continuum along which natural selection can operate, then such adaptive goals become inevitable.

One of the stratagems adopted by Darwin in the *Origin* and used by many evolutionary biologists since, when faced with the difficulty of envisaging transitional forms, is to allude to the poverty of the human imagination and to the very surprising and curious adaptations and behaviour patterns many organisms exhibit – the implication being that had we not known of such bizarre adaptations we

would never have believed them possible. Shortly after his discussion
of bat evolution he makes the point:[47]

> If about a dozen genera of birds were to become extinct, who would
> have ventured to surmise that birds might have existed which used
> their wings solely as flappers, like the logger headed duck (Micro-
> pterus of Eyton); as fins in the water and as front-legs on land, like the
> penguin; as sails, like the ostrich; and functionally for no purpose, like
> the *Apteryx*? Yet the structure of each of these birds is good for it,
> under the conditions of life to which it is exposed, for each has to live
> by a struggle; but it is not necessarily the best possible under all poss-
> ible conditions. It must not be inferred from these remarks that any of
> the grades of wing-structure here alluded to, which perhaps may all be
> the result of disuse, indicate the steps by which birds actually acquired
> their perfect power of flight; but they serve to show what diversified
> means of transition are at least possible.
>
> In North America the black bear was seen by Hearne swimming for
> hours with widely open mouth, thus catching, almost like a whale,
> insects in the water.[48]

In effect, what Darwin is saying, and what many subsequent
evolutionists have echoed, is that though we cannot imagine exactly
how the gaps were bridged in any particular case this is merely
because our imagination is relatively crude alongside the ingenuity of
nature. Thus the problem of providing detailed reconstructions of
credible sequences of transitional forms is avoided and we are asked
instead to wonder at the bountiful creativity of nature. But rather
than convince, this strategy only tends to emphasize the fundamental
inability of evolutionary theorists to confront the problem of the
gaps. Further, this sort of argument smacks of tautology. Of course,
if gradual evolution is true then the gaps must have been closed
gradually even if we can't imagine how it occurred!

Until recently, despite the severity of the problem of reconstructing transitional stages, few biologists have been prepared
to reject gradualism altogether. However over the past few years
a number of biologists and students of evolution theory have
begun to raise serious doubts about the validity of orthodox Darwinian gradualism. As Stephen Jay Gould put it in *The Panda's
Thumb*:[49]

> . . . . can we invent a reasonable sequence of intermediate forms – that
> is, viable, functioning organisms – between ancestors and descendants

in major structural transitions? I submit, although it may only reflect my lack of imagination, that the answer is no . . .

And the same sentiment was expressed by Stanley in his book *Macroevolution*.[50] Many other authorities have also recently expressed scepticism over the gradualistic philosophy, enshrined in the views of Heilman and Ostrom, arguing that it cannot account for the major innovations of evolution in a fully plausible and comprehensive manner.

Ultimately there is, of course, absolutely no reason why functional organic systems should form the continuum that evolution by natural selection demands. In the world of physics and chemistry many phenomena are discontinuous. One cannot gradually convert one molecular species into another, neither can one convert gradually one type of atom into another. Between such entities there are jumps. Might not functional organic systems be similarly separated by discontinuities? The sentiment was expressed by the Scots zoologist D'Arcy Thompson:[51]

An algebraic curve has its fundamental formula, which defines the family to which it belongs: . . . We never think of "transforming" a helicoid into an ellipsoid or a circle into a frequency-curve. So it is with the forms of animals. We cannot transform an invertebrate into a vertebrate, nor a coelenterate into a worm, by any simple and legitimate deformation, nor by anything short of reduction to elementary principles. . . . The lines of the spectrum, the six families of crystals, the chemical elements themselves, all illustrate this principle of discontinuity. In short nature proceeds from one type to another among organic as well as inorganic forms; and these types vary according to their own parameters, and are defined by physico-mathematical conditions of possibility. In natural history Cuvier's "types" may not be perfectly chosen but types they are; and to seek for stepping-stones across the gaps between is to seek in vain, for ever.

If the divisions of nature really are as fundamental as Cuvier insisted and cannot be crossed gradually through a series of functional transitional forms, then the only alternative is to conceive of evolution in terms of a succession of frozen accidents whereby wholly new organs, types and adaptations suddenly emerge as a result of some sort of fortuitous macromutational event.

Scepticism of gradualism obviously leads to the idea of evolution by saltation, which was the path taken by Goldschmidt in his *The Material Basis of Evolution* in 1940, where he introduced the notion of the "hopeful monster" as a means of getting from one type to another suddenly in one jump:[52]

> A monstrosity appearing in a single genetic step might permit the occupation of a new environmental niche and thus produce a new type in one step. A Manx cat with a hereditary concrescence of the tail vertebrae, or a comparable mouse or rat mutant, is just a monster. But a mutant of Archaeopteryx producing the same monstrosity was a hopeful monster because the resulting fanlike arrangement of the tail feathers was a great improvement in the mechanics of flying.

Frustrated with the empirical absence of intermediate forms and with the difficulty of conceiving of gradual functional transitions, there has been an upsurge recently of this traditional alternative to gradualism, the concept of evolution by saltation, the idea that new organs and types emerge suddenly following some sort of massive macromutation.

As we have seen, Darwin considered that the sudden appearance of a new adaptive structure or organ would be a miracle[53] and this has been the position taken by the great majority of biologists ever since. Mayr comments on the possibility of a "hopeful monster":[54]

> The occurrence of genetic monstrosities by mutation . . . is well substantiated, but they are such evident freaks that these monsters can be designated only as "hopeless". They are so utterly unbalanced that they would not have the slightest chance of escaping elimination through selection. Giving a thrush the wings of a falcon does not make it a better flyer. Indeed, having all the other equipment of a thrush, it would probably hardly be able to fly at all. . . . To believe that such a drastic mutation would produce a viable new type, capable of occupying a new adaptive zone, is equivalent to believing in miracles.

While it might be theoretically possible to avoid the impasse of gradualism by opting for saltation, it seems unlikely that purely random processes would ever throw together suddenly adaptations like a feather or the avian lung or the amniotic egg. The likely improbability of evolution by "saltation" is the subject of Chapter Thirteen.

NOTES

1. Thieme, P. (1958) "The Indo-European Language", *Scientific American*, 199 (4), pp 63–74.
2. Heilman, G. (1926) *The Origin of Birds*, Witherby, London, from Figure 142, p 199.
3. Rensch, B. (1959) Evolution above the Species Level, Columbia University Press, New York, p 166.
4. Tyne, Van T., and Berger, A. J. (1959) *Fundamentals of Ornithology*, John Wiley and Sons, New York.
5. Ostrom, J. H. (1979) "Bird Flight: How Did It Begin?", *American Scientists*, 67: 45–56, p 46.
6. Heilman, op cit, pp 200–01.
7. Caple, G., Balda, R. P., and Willis, W. R. (1983) "The Physics of Leaping Animals and the Evolution of Preflight", *Amer. Nat.*, 121: 455–67, see p 474
8. Ostrom, op cit, p 47.
9. ibid, p 47.
10. ibid, p 47.
11. ibid, p 47.
12. ibid, p 47.
13. Caple, op cit.
14. Heilman, op cit.
15. Ostrom, op cit, p 55.
16. ibid, from Figure 10, p 55.
17. Stahl, B. J. (1974) *Vertebrate History. Problems in Evolution*, McGraw-Hill Book Co, New York, p 349.
18. Duncker, H. R. (1978) "Development of the Avian Respiratory and Circulation Systems", in *Respiratory Function in Birds, Adult and Embryonic*, ed J. Piiper, Springer Verlag, New York, pp 260–73, see Figure on p 269.
19. Duncker, op cit.
20. Cuvier, G. (1800) cited in Coleman, W. (1964) *Georges Cuvier: Zoologist*, Harvard University Press, Cambridge, Mass, pp 171–72.
21. Darwin, C. (1872) The Origin of Species, 6th ed (1962) Collier Books, New York, p 182.
22. ibid, p 173.
23. Jepsen, G. L. (1970) "Bat Origins and Evolution" in *Biology of Bats*, 2 vols, vol 1, ed W. A. Wimsatt, pp 1–64, see pp 42, 46.
24. ibid, p 44.
25. Smith, J. D. (1976) "Comments on Flight and Evolution of Bats" in *Major Patterns of Evolution*, ed M. R. Hecht, Plenum Press, London, from Figure 2, p 429.
26. ibid, pp 427–30.
27. ibid, from Figure 1, p 428.
28. Jepsen, op cit, p 53.
29. Thomson, J. A. (1934) *Biology for Everyman*, 2 vols, J. M. Dent and Sons, vol 1, p 682.
30. Dewar, D. (1938) *More Difficulties of the Evolution Theory*, Thynne and Co, London, pp 23–4.

31. Tillyard, R. J. (1917) *The Biology of the Dragonfly*, Cambridge University Press, Cambridge, p 215.
32. Chandler, A. C. (1961) *Introduction to Parasitology*, 10th ed, J. Wiley and Sons, New York, p 16.
33. Bertalanffy, L. (1969) "Chance or Law", in *Beyond Reductionism*, ed A. Koestler, Hutchinson Publishing Co, London, pp 56–84, see p 65.
34. Hardin, G. (1968) *39 Steps to Biology*, W. H. Freeman and Co, San Francisco, p 105.
35. Bristow, W. S. (1958) *The World of Spiders*, Collins, London, p 177.
36. Petrunkevitch, A. (1952) "The Spider and the Wasp", *Scientific American*, 187 (2): 20–23, p 21.
37. Crystal, R. N. (1930) *Studies of the Sirex Parasites*, Oxford University Press, Oxford, p 46.
38. Berg, H. (1975) "How Bacteria Swim", *Scientific American*, 233 (2): 36–44, from Figure on p 44.
39. ibid, p 44.
40. Darwin, C. (1882) *The Various Contrivances by which Orchids Are Fertilised by Insects*, 2nd ed, John Murray, London, p 173.
41. ibid, pp 173–74.
42. Lloyd, F. E. (1942) *The Carnivorous Plants*, Chronica Botanical Co, Waltham, Mass, p 7.
43. Wardlaw, C. W. (1965) *Organization and Evolution in Plants*, Longmans, Green and Co Ltd, London, p 405.
44. ibid, p 405.
45. Huxley, J. (1942) Evolution: *The Modern Synthesis*, Allen and Unwin, London, pp 413–14.
46. Wardlaw, op cit, p 405.
47. Darwin, op cit, p 174.
48. ibid, p 176.
49. Gould, S. J. (1980) *The Panda's Thumb*, W. W. Norton and Co, New York, p 189.
50. Stanley, S. (1980) *Macroevolution*, W. H. Freeman and Co, San Francisco.
51. Thompson, D'Arcy W. (1942) *On Growth and Form*, 2nd ed, 2 vols, Cambridge University Press, Cambridge, vol 2, p 1094.
52. Goldschmidt, R. (1940) *The Material Basis of Evolution*, Yale University Press, New Haven, p 390.
53. Darwin, op cit, p 242.
54 Mayr, E. (1970) *Populations, Species and Evolution*, Harvard University Press, Cambridge, Mass, p 253.

CHAPTER 10
The Molecular Biological Revolution

The laws governing inheritance are for the most part unknown. No one can say why the same peculiarity in different individuals of the species is sometimes inherited and sometimes not so . . . why a peculiarity is often transmitted from one sex to both sexes or to one sex alone.

It is difficult to think of a comparable decade in scientific history when fundamental knowledge increased as quickly as it did in biochemistry in the 1950s. Before 1950 hardly anything was known of the molecular basis of life; yet during the next ten years a succession of dramatic discoveries completely transformed the biological sciences and laid the foundation for a totally new description of life.

In 1953 Watson and Crick published their now famous paper in the journal *Nature* reporting the double helical structure of a then obscure compound deoxyribonucleic acid – DNA. This momentous discovery solved the centuries-old puzzle of heredity, revealing its chemical basis and turning the term "double helix" into a household word. Just two years later Sanger reported the first complete chemical structure of a protein, insulin, an achievement which had taken him ten years of meticulous work. In 1957 x-ray crystallographic studies of sperm whale myoglobin provided the first picture of the 3-D structure of a protein, and in 1959 Perutz announced the 3-D structure of horse oxyhaemoglobin. Over the next few years rapid advances in many different areas of biochemistry began to reveal the structure and function of all the main molecular components of the cell.

The biochemical knowledge about living systems that has accumulated over the past twenty years has provided a vast new body of information by which to assess evolutionary claims and as a result a number of interesting problems have arisen. Some of the more important ones are discussed in the next four chapters. So that the reader unfamiliar with molecular biology will be able to grasp the main points and arguments presented this chapter provides a brief introductory description of the basic chemical structure and function of some of the key molecules of life such as the proteins and nucleic acids and of the role they play in the cell.

Protein molecules are the ultimate stuff of life. If we think of the cell as being analogous to a factory, then the proteins can be thought of as analogous to the machines on the factory floor which carry out individually or in groups all the essential activities on which the life of the cell depends. Each protein is a sort of micro-miniaturized machine, so small that it must be magnified a million times before it is visible to the human eye. The structure and functioning of these fascinating work horses of the cell was a complete mystery until the 1950s.

We now know that each of these tiny molecular machines consists fundamentally of a long chain-like molecule, or polymer, made up of a linear sequence of simple organic compounds called amino acids. Of the hundreds of amino acids known to science only twenty are utilised by living systems in the construction of proteins.

Amino acid symbols

Amino acid	Three-letter symbol	Amino acid	Three-letter symbol
Alanine	Ala	Isoleucine	Ile
Arginine	Arg	Leucine	Leu
Asparagine	Asn	Lysine	Lys
Aspartic acid	Asp	Methionine	Met
Asn and/or Asp	Asx	Phenylalanine	Phe
Cysteine	Cys	Proline	Pro
Glutamine	Gln	Serine	Ser
Glutamic acid	Glu	Threonine	Thr
Gln and/or Glu	Glx	Tryptophan	Trp
Glycine	Gly	Tyrosine	Tyr
Histidine	His	Valine	Val

Table giving amino acids used in proteins and the three letter symbols used in scientific literature

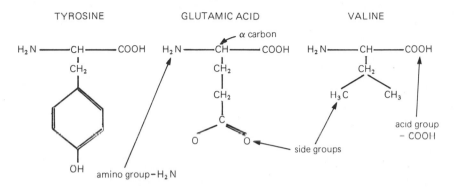

Figure 10.1: The Chemical Structure of Three Amino Acids.

Amino acids are small organic compounds consisting of about ten to twenty atoms. The structure of three amino acids is shown in Figure 10.1

Each amino acid contains an amino (NH_2) and a carboxyl acid (COOH) group linked by a carbon atom, as well as a unique side chain. Because of their different side chains each amino acid has different chemical properties. Some are insoluble in water (hydrophobic), some are soluble (hydrophilic), some are basic while others are acidic.

The amino acids are linked via their amino and carboxyl acid groups to form a long linear polymer which is known as the primary structure of the protein. Figure 10.2 shows the chemical structure of a section of an amino acid chain. The backbone which is formed by the linkage of the amino acid carboxylic acid groups is identical throughout the molecule. It is the unique side groups which jut out from the backbone which confer different chemical properties to different regions of the amino acid chain.

The linear sequence of amino acids in a protein can be thought of as a sentence made up of a long combination of the twenty amino acid letters. Just as different sentences are made up of different sequences of letters, so different proteins are made up of different sequences of amino acids. In most proteins the amino acid chain is between one hundred and five hundred amino acids long.

Every different protein has a unique amino acid sequence and this is known as its primary structure. The figure below[1] gives the

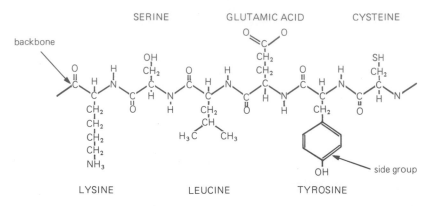

Figure 10.2: The Chemical Structure of a Short Section of the Amino Chain of a Protein.

primary structure of the small protein ribonuclease (the standard three letter abbreviations are used to indicate the different amino acids in the chain):

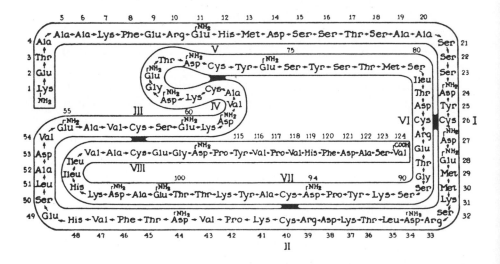

The linear chain of amino acids folds automatically under the influence of various electro-chemical forces into a complex 3-D aggregate of atoms referred to as its tertiary structure (see Figure 10.3). Chain folding occurs in such a way so as to bring about the maximum number of favourable atomic interactions between the various constituent amino acids, and, during it, which takes in the order of a second, negatively charged groups tend to associate with positively charged groups and amino acids with hydrophobic side chains tend to stack in the centre of the molecule while amino acids with hydrophilic side chains tend to arrange themselves on the surface in contact with water. The final stable tertiary shape, also known as the minimum energy conformation, is dictated directly by the amino acid sequence.

Specific amino acid sequences lead to specific 3-D shapes. However, although the final 3-D conformation is determined directly by the amino acid sequence, minor changes in the amino acid sequence of a very conservative nature can still lead to the same basic 3-D conformation in the tertiary structure of the molecule. For example, the

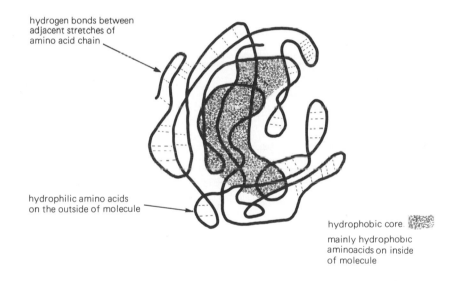

hydrogen bonds between adjacent stretches of amino acid chain

hydrophilic amino acids on the outside of molecule

hydrophobic core.
mainly hydrophobic aminoacids on inside of molecule

Figure 10.3: The Structural Organization of a Protein. *The final 3D conformation of the amino acid chain is held in a stable conformation by a variety of weak chemical interactions. These include the hydrogen bonds which hold together amino acids in adjacent coils and the hydrophobic bonds which hold together in the centre of the molecule a cluster of amino acids which are "water avoiding", being relatively insoluble and unable to undergo electrostatic interactions with water. This cluster forms the stable core of the molecule around which the outer regions of the amino acid backbone are entwined. On the outside of the molecule are the hydrophilic or soluble amino acids which are able to undergo weak electrostatic interactions with the water molecules around the outside of the molecule.*

substitution of a hydrophobic amino acid for another hydrophobic amino acid may have no effect on the formation of a stable hydrophobic core and hence no influence on the final 3-D form of the molecule. But if a hydrophilic amino acid is substituted for a hydrophobic one in the centre of the molecule then the attraction of the hydrophilic amino acid for the water at the surface of the protein may destabilise the entire molecule.

Most proteins consist altogether of some several thousand atoms folded into an immensely complex spatial arrangement. Proteins which perform different functions have completely different overall 3-D structures and functional properties. The proteins which form the keratin in hair and nails are long and thin and intertwined round each other like the fibres in a length of wool. Other proteins known as enzymes, those which carry out particular chemical reactions, tend to be rounded in shape and possess a special region in the molecule known as the *active site*. This is generally a cleft-like structure which extends from the surface of the molecule into its interior and into which fits precisely the compound upon which the enzyme acts which is known as the substrate.

Apart from structural and catalytic functions, proteins also carry out transport and logistic functions and, because of the enormous number of different protein functions, the variety of different sorts and shapes of proteins is correspondingly very great.

Although proteins are amazingly versatile and carry out all manner of diverse biochemical functions they are incapable of assembling themselves without the assistance of another very important class of molecules – the nucleic acids. To return again to the analogy of the factory, while the proteins can be thought of as the working elements of a factory, the nucleic acid molecules can be thought of as playing the role of the library or memory bank containing all the information necessary for the construction of all the various machines (proteins) on the factory floor. More specifically, we can think of the nucleic acids as a series of blueprints, each one containing the specification for the construction of a particular protein in the cell.

There are two types of nucleic acids, DNA and RNA. DNA is only found in the nucleus of the cell, equivalent to the head office of the factory, and contains the master blueprints. RNA molecules perform the fundamental task of carrying the information stored in DNA to all the various parts of the cell where the manufacture of a particular protein is proceeding. In terms of our analogy we can think of RNA molecules as photocopies of the master blueprint (DNA) which are carried to the factory floor where the technicians and engineers convert the abstract information of the blueprint (RNA) into the concrete form of the machine (protein).

So, according to what is known as the fundamental dogma of molecular biology, information in living systems travels from:

DNA   (master blueprint)
↓ ← – – transcription
RNA   (photocopy)
↓ ← – – translation
PROTEINS   (functional machines)

Just as in the factory, the information in the blueprint flows via the photocopy into a manufactured article on the factory floor.

In terms of their actual structure, of course, nucleic acid molecules do not resemble blueprints but are long chain-like molecules. If we wish to continue thinking in terms of the factory/cell analogy a better picture of the nucleic acids would be to think of them as analogous to magnetic tapes which are often used nowadays to programme automatic lathes or jigborers in the production of machine tools.

Both DNA and RNA are similar therefore to proteins in one aspect; they are long chain-like molecules formed by the linking together of small subunits. The building blocks or subunits of DNA polymers are the nucleotides, each consisting of a phosphate radical, a sugar, and a nitrogenous base. There are four nucleotides in DNA and these are linked together to form a long linear polymer. In the great majority of living things DNA exists in a stable double stranded conformation in which the two strands are held together by specific chemical associations – hydrogen bonds between the bases in the two strands. For chemical reasons the adenine (A) in one strand is associated with thymine (T) in the other, while guanine (G) is associated with cytosine (C) (see Figure 10.4). The two strands are twisted round each other to form the so-called double helix.

Unlike DNA, RNA is a single stranded polymer made up of four nucleotides of very similar chemical structure to those of DNA. The only difference is that one of the nucleotides of RNA contains the base uracil instead of thymine but, as uracil and thymine have similar chemical properties, this makes little difference. (For example, uracil can form a chemical association with adenine in double stranded nucleic acids.) The nucleotides of RNA are again linked via their phosphate radicals to form a long polymer similar to a single strand of DNA.

Unlike proteins, nucleic acids do not fold up into complex 3-D conformations but remain as relatively simple long chain-like objects. Some DNA molecules may consist of several million subunits and when fully extended stretch for several centimetres.

The linear sequence of subunits in the DNA molecule contains a series of encoded messages, genes, each of which is decoded by the cell and translated into the linear sequence of amino acids of a protein.

Although the sequence of nucleotides in the DNA of the gene is the ultimate store of information necessary for the specification of the

Figure 10.4: The Structure of DNA. *DNA consists of a long sequence of compounds called nucleotides. Each nucleotide consists of a sugar, a nitrogen containing base and a phosphate group. The nucleotides in a DNA chain are linked together through their phosphate groups. Four nucleotides occur in DNA and they only differ in the structure of their constituent nitrogen containing base. These four bases adenine, guanine, cytosine, and thymine, are usually abbreviated as A, G, C, and T. They are the letters of the genetic alphabet. For chemical reasons the adenine (A) in DNA tends to form a spontaneous association with thymine (T) while guanine (G) associates with cytosine (C). DNA is made up of two long strands joined by means of these specific associations. In the double strand, A of one strand is linked to T in the other, G to C, T to A, and C to G. The two strands are therefore complementary. This structure is represented in the figure opposite, the pentagons represent sugar residues, the black circles – phosphate groups and the shaded shapes marked A, T, G, and C – the bases matched in pairs.* (from Monod)[2]

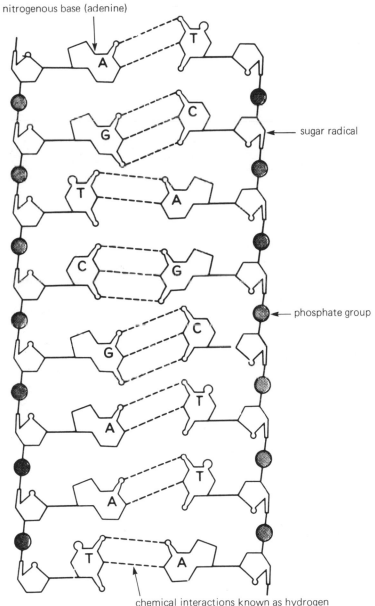

nitrogenous base (adenine)

sugar radical

phosphate group

chemical interactions known as hydrogen bonds between complementary bases in the two DNA strands

amino acid sequence of a protein, as mentioned above the nucleotide sequence of the DNA itself is not read directly into the amino acid sequence of a protein. Rather the nucleotide sequence of the DNA is first copied into the nucleotide sequence of a particular type of RNA known logically as messenger RNA (mRNA). The process of copying the nucleotide sequence of the gene is known as transcription.

At transcription one of the two strands of the DNA double helix is copied into RNA. Firstly, the helix is unwound and, secondly, one of the strands directs the synthesis of an RNA polymer of complementary nucleotide sequence. This can be seen in the figure below. The transcription of mRNA is carried out by a complex of proteins known as RNA polymerase.

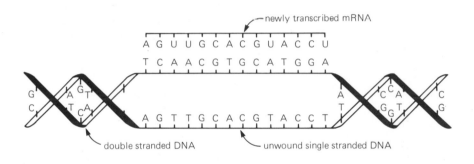

As most genes are about one thousand nucleotides long (this being the length of DNA necessary to specify for the average protein), each mRNA molecule, being merely a copy of a gene, consists of a long RNA chain about one thousand nucleotides in length.

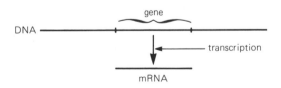

The above description of the synthesis of the mRNA molecule applies mainly to the process as it occurs in bacterial cells. The situation is somewhat more complicated in higher organisms because the coding sequences are separated by intervening sequences, or introns. After the transcription of the DNA, the initial RNA transcript is subjected to processing during which the nutrons are removed and the remaining coding sequences spliced together to form the mature mRNA molecule.

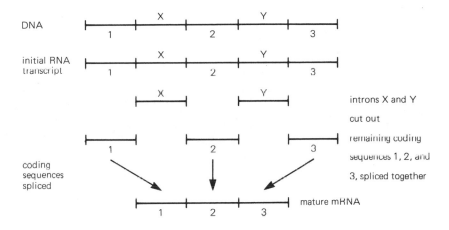

The sequence of nucleotides in the mRNA is translated by the conventions of the genetic code into the amino sequence of a protein in the same way as a message in Morse code can be translated into a sequence of letters by applying the translational conventions of Morse. Hence a sequence of dots and dashes

$$-- .- -.$$

can be decoded according to the rules of Morse where

$$-- = M$$
$$.- = A$$
$$-. = N$$

to give the sequence of letters MAN.

In precisely the same way, a sequence of nucleotides in an RNA molecule, for example

$$A G U \ C G A \ U U G \ A C A$$

can be translated by applying the following rules of the genetic code:

where   AGU = the amino acid serine (SER)
     CGA = the amino acid orginine (ARG)
    UUG = the amino acid leucine (LEU)
    ACA = the amino acid threnine (THR)

into the amino acid sequence

SER – ARG – LEU – THR

The nucleotide sequence in mRNA is read in successive non-overlapping triplets such that successive triplets of nucleotides in the mRNA specify successive amino acids in the protein. Every one of the sixty-four different nucleotide triplets which can be formed from the four nucleotides, A, U, G, C, has an exact meaning. Sixty-one triplets specify for amino acids. The remaining three which happen to be the triplets – UAA, UAG, UGA – are used as punctuation signals and mean "Stop" indicating the end of a particular message.

Two triplets have a double meaning and, depending on their position in the mRNA molecule and the surrounding nucleotide sequence, can also act as "Start" signals. These are the triplets AUG and GUG. AUG sometimes means the amino acid methionine and at other times means "Start" while GUG may mean the amino acid valine or "Start".

One of the differences between the Morse and genetic codes is that while in Morse there is only one unique combination of dots and dashes for each letter, in the case of the genetic code there are several different nucleotide triplets for some amino acids, for example the

UUU		UCU		UAU		UGU	
UUC	Phenylalanine	UCC	Serine	UAC	Tyrosine	UGC	Cysteine
UUA		UCA		UAA		UGA	Stop
UUG		UCG		UAG	Stop	UGG	Tryptophan
CUU	Leucine	CCU		CAU	Histidine	CGU	
CUC		CCC	Proline	CAC		CGC	Arginine
CUA		CCA		CAA	Glutamine	CGA	
CUG		CCG		CAG		CGG	
AUU		ACU		AAU	Asparagine	AGU	Serine
AUC	Isoleucine	ACC	Threonine	AAC		AGC	
AUA		ACA		AAA	Lysine	AGA	Arginine
AUG	Methionine or start	ACG		AAG		AGG	
GUU	Valine	GCU		GAU	Aspartic acid	GGU	
GUC		GCC	Alanine	GAC		GGC	Glycine
GUA	Valine or start	GCA		GAA	Glutamic acid	GGA	
GUG		GCG		GAG		GGG	

triplets UCG, UCA, UCU, UCC, AGU, AGC, all code for the one amino acid serine. The genetic code is therefore redundant.

After its transcription the mRNA moves from the nucleus into the cytoplasm to the actual site of translation where the decoding of the message takes place. The translation of the mRNA molecule is carried out by a complex set of molecules which together constitute the translational apparatus. An important component of the translational apparatus is a complex globular organelle, known as the *ribosome,* composed of an aggregate of some 50 proteins and three chains of RNA. The ribosome attaches itself to the mRNA at a special site on the mRNA known as the ribosome binding site which contains a "Start" triplet AUG or GUG (see table opposite). This site is generally close to one end of the mRNA molecule. Like any other automatic decoding system the translational system in the cell includes a set of transducing elements which relate each functional unit of the code, ie each triplet in the RNA, to the correct item in the translated message, that is, a particular amino acid. This key function is carried out by a special class of RNA molecules known as transfer or tRNA. Each tRNA molecule consists of a short polymer of RNA some one hundred nucleotides long folded into a compact hairpin looped structure. Each tRNA can recognize a particular triplet in the mRNA as well as the appropriate amino acid specified according to the conventions of the code.

During the process of translation the mRNA passes through the ribosome just as the magnetic tape passes the recording head on a tape recorder. As each triplet reaches the reading head, it associates loosely with its appropriate tRNA which is also carrying the appropriate amino acid. Special proteins in the ribosome remove the amino acid from the tRNA and hence the amino acid chain is gradually assembled amino acid by amino acid as successive tRNA bring their attached amino acids to the reading head of the ribosome (see Figure 10.5). When the amino acid chain is completed it is detached from the tRNA and folds automatically into its correct 3-D functional conformation.

The synthesis of proteins by the cell is thus achieved as a result of a remarkable and intimate relationship between one class of molecules – the proteins – and another quite different class of molecules – the nucleic acids. The nucleic acids contain the information for the construction of proteins, but it is the proteins which extract and utilize that information at all stages as it flows through this intricate series of transformations.

Another important process which goes on in the cell is that of DNA

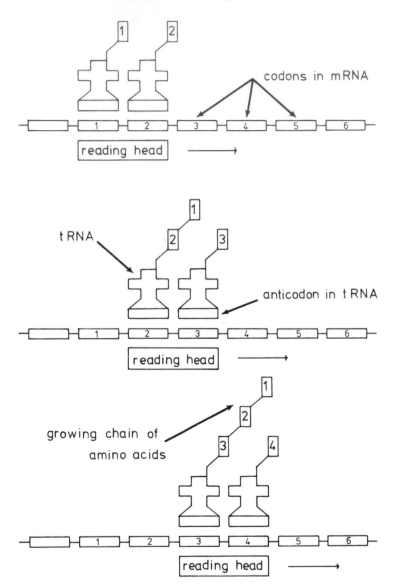

Figure 10.5: The Elongation of the Amino Acid Chain. *The ribosome moves along the mRNA. As a codon passes the reading head the corresponding tRNA molecule with appropriate amino acid attached associates briefly with the correct codon in the mRNA (via base pairing between codon and anticodon). The ribosome attaches the amino acid or amino acid chain of the preceding tRNA onto the amino acid of the next incoming tRNA and hence the amino acid chain is gradually elongated.*

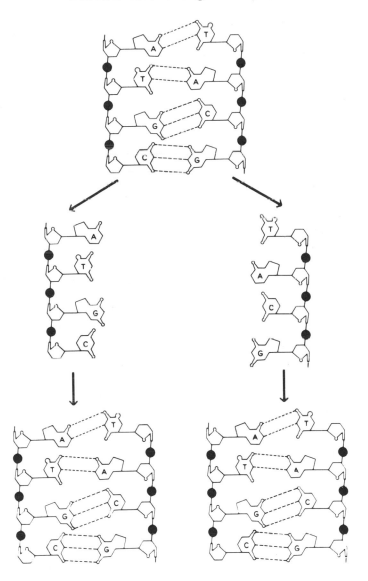

Figure 10.6: DNA Replication. *The replication of this molecule proceeds by the separation of the duplex, followed by the reconstruction nucleotide by nucleotide, of the two complements. This is shown in a simplified manner and confining ourselves to four base pairs. The two molecules thus synthesized each contain one strand of the parent molecule and a strand newly formed by specific nucleotide-by-nucleotide pairing. The two new molecules are identical to each other and to the original molecule. Thus DNA is replicated.* (from Monod)[3]

replication. Although its inherent chemical structure greatly facilitates its replication (see Figure 10.6), the process still depends on the activities of a number of proteins. Because DNA can encode for all the proteins necessary for transcription, translation and *its own replication*, the cell system can replicate itself. The figure below summarizes the main interactions between nucleic acids and proteins in the cell:

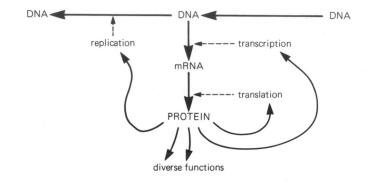

NOTES

1. Smyth, D. G., Stein, W. H. and Moore, S. (1963) "The Sequence of Amino Acid Residues in Bovine Ribonuclease", *Journal Biol. Chem*, 238:227–234, see Figure 1, p 228.
2. Monod, J. (1972) *Chance and Necessity*, Collins, London, see Figure on p 178.
3. ibid, Figure on 179.

CHAPTER 11

The Enigma of Life's Origin

It is often said that all the conditions for the first production of a living organism are now present, which could ever have been present. But if (and oh! what a big if!) we could conceive in some warm little pond, with all sorts of ammonia and phosphoric salts, light, heat, electricity, etc, present, that a protein compound was chemically formed ready to undergo still more complex changes, at the present day such matter would be instantly devoured or absorbed, which would not have been the case before living creatures were formed.

Even as recently as the nineteenth century when all the major morphological discontinuities of nature had been described and classified, the question as to whether there was a distinct break between life and the inorganic world was controversial. The existence of a definite discontinuity was only finally established after the revolutionary discoveries of molecular biology in the early 1950s. Before then it was still possible to hope that perhaps advances in science would reveal a number of intermediates between chemistry and the cell. The possibility existed, for example, that certain viruses would prove to be intermediate between the physical and biological worlds. The hope that increased biochemical knowledge would bridge the gap was specifically expressed by many authorities in the 1920s and 30s. But, as in so many other fields of biology, the search for continuity, for empirical entities to bridge the divisions of nature, proved futile. Instead of revealing a multitude of transitional forms through which the evolution of the cell might have occurred, molecular biology has served only to emphasize the enormity of the gap.

We now know not only of the existence of a break between the living and non-living world, but also that it represents the most dramatic and fundamental of all the discontinuities of nature. Between a living cell and the most highly ordered non-biological system, such as a crystal or a snowflake, there is a chasm as vast and absolute as it is possible to conceive.

Molecular biology has shown that even the simplest of all living systems on earth today, bacterial cells, are exceedingly complex objects. Although the tiniest bacterial cells are incredibly small, weighing less than 10^{-12}gms, each is in effect a veritable micro-miniaturized factory containing thousands of exquisitely designed pieces of intricate molecular machinery, made up altogether of one hundred thousand million atoms, far more complicated than any machine built by man and absolutely without parallel in the non-living world.

Molecular biology has also shown that the basic design of the cell system is essentially the same in all living systems on earth from bacteria to mammals. In all organisms the roles of DNA, mRNA and protein are identical. The meaning of the genetic code is also virtually identical in all cells. The size, structure and component design of the protein synthetic machinery is practically the same in all cells. In terms of their basic biochemical design, therefore no living system can be thought of as being primitive or ancestral with respect to any other system, nor is there the slightest empirical hint of an evolutionary sequence among all the incredibly diverse cells on earth. For those who hoped that molecular biology might bridge the gulf between chemistry and biochemistry, the revelation was profoundly disappointing.

In the words of Monod:[1]

> . . . we have no idea what the structure of a primitive cell might have been. The simplest living system known to us, the bacterial cell . . . in . . . its overall chemical plan is the same as that of all other living beings. It employs the same genetic code and the same mechanism of translation as do, for example, human cells. Thus the simplest cells available to us for study have nothing "primitive" about them. . . no vestiges of truly primitive structures are discernible.

We have already seen that in the case of the great morphological divisions, where empirical evidence of intermediates is lacking, there is invariably a conceptual problem in envisaging fully plausible hypothetical intermediates through which evolution could have occurred. As we shall see in this chapter precisely the same sort of conceptual problem is met in trying to reconstruct the hypothetical sequence of transitional systems which led eventually to the modern cell.

In the *Origin* Darwin made no claim that his model of evolution could be extended to explain the origin of life, but the implication was there and was soon taken up by some of his contemporaries like Thomas Huxley. Huxley speculated:[2]

> Looking back through the prodigious vista of the past, I find no record of the commencement of life, and therefore I am devoid of any means of forming a definite conclusion . . . but . . . if it were given to me to look beyond the abyss of geologically recorded time to the still more remote period when the Earth was passing through physical and chemical conditions which it can no more see again than a man can recall his infancy, I should expect to be a witness of the evolution of living protoplasm from not-living matter.

It has become almost axiomatic today among evolutionary biologists that the same gradual process which drove the evolution of life, the successive selection of beneficial mutations, was also responsible for its creation. Accordingly, the first cell is supposed to have arisen following a long period of pre-cellular evolution. The process is presumed to have begun with a primitive self-replicating molecule which slowly accumulated beneficial mutations that enabled it to reproduce more efficiently. After eons of time, it gradually evolved into a more complex self-replicating object acquiring a cell membrane, metabolic functions and eventually all the complex biochemical machinery of the cell. As the outcome of a perfectly natural process, driven only by chance and selection, life is now widely viewed as an inevitable product of any planetary surface which has the correct geochemical and geophysical character.

The depth to which our culture now adheres to this idea explains the current belief that life is widespread in the universe – why serious attempts have been made to detect messages sent to earth by extraterrestrial civilizations and why NASA placed a plaque giving information about life on earth on the spacecraft *Pioneer 10*. It also explains why so much energy was recently devoted to the search for life in the Martian soil.

The discovery of life, especially if it were to prove widespread, would of course have a very important bearing on the question of how life originated on earth. For it would undoubtedly provide powerful circumstantial evidence for the traditional evolutionary scenario, enhancing enormously the credibility of the belief that the route from chemistry to life can be surmounted by simple natural processes wherever the right conditions exist. In this context the recent growth of interest in exobiology is perfectly understandable. As American astronomer Carl Sagan remarks in his book *Intelligent Life in the Universe*:[3]

> . . . the discovery of life on one other planet – e.g. Mars – can, in the words of the American physicist Philip Morrison, of the Massachusetts Institute of Technology, "transform the origin of life from a miracle to a statistic."

The possibility of life on other worlds has intrigued man for centuries. In the seventeenth century the Dutch physicist Christianus Huygens wrote a book entitled *New Conjectures Concerning the Planetary Worlds their Inhabitants and Productions* in which he freely speculated about the possibility of extraterrestrial life:[4]

> And yet 'tis not improbable that those great and noble Bodies have somewhat or other growing and living upon them, though very different from what we see and enjoy here. Perhaps their Plants and Animals may have another sort of Nourishment there.

And not just life but intelligent rational beings:[5]

> That which makes me of this Opinion, that those Worlds are not without such a Creature endued with Reason, is that otherwise our Earth would have too much the Advantage of them, in being the only part of the Universe that could boast of such a Creature . . .

In the nineteenth century one of Darwin's own mentors at Cambridge, the Reverend William Whewell, was another believer in extraterrestrial life, although not in life forms as intelligent as man. In the *Plurality of Worlds*, he wrote of the inhabitants of Jupiter:[6]

> Who can conceive the configuration of the creatures that dwell there? They may exist as immense algae-like or medusa-like creatures, "floating many a rood", . . . they must . . . it would seem, be cartilaginous

and glutinous masses. If life be there, it does not seem in any way likely that the living things can be anything higher in the scale of being than such boneless watery, pulpy creatures . . .

In the twentieth century, and especially over the past few decades, the idea that there is not only life in space but also civilizations greatly in advance of our own has become commonplace. So widespread has this belief become in scientific circles that since the early 60s a number of radioastronomers have set up programmes to search regions of the sky for intelligent signals. The most publicized was that nicknamed "Project OZMA", set up by the American radio-astronomer Frank Drake who worked at the time at the National Radio Astronomy observatory in Green Bank, West Virginia. "Project OZMA" was relatively limited and only two hundred hours were spent listening to two nearby stars for radio signals.

Of all the planets in our solar system, the red planet Mars has always been traditionally cited as the most likely abode for extraterrestrial life. This is not only because its atmosphere and climate are more like Earth's than any other planet, but also because of various peculiar surface features which have been observed by astronomers on Earth.

Nearly everyone has heard of the Martian canals. These were first observed by the Italian astronomer, Giovanni Schiaparelli, in 1877. Schiaparelli did not claim to see canals in the English sense of the word, which carries the distinct implication of intelligent design. He had merely observed long straight lines which appeared to connect the dark areas on the Martian surface and used the word "canali", which is Italian for channels or grooves. But some years later the American astronomer, Percival Lowell, having observed lines on Mars himself, vigorously championed the possibility that the "canali" of Schiaparelli were indeed canals – the artifacts of a civilized race. Lowell made a daring series of deductions based on the "canals". Carl Sagan, himself a strong believer in extraterrestrial life, describes the lengths to which Lowell went:[7]

Lowell and his followers constructed an inverted pyramid of deductions upon the apex of the canal observations. The canals were a massive engineering work; therefore, the Martians are in substantial technological advance of contemporary human society. The canals obviously cross what we would term international boundaries; hence, a world government exists on Mars. One of Lowell's followers went so far as to place the capital in Solis Lacus (latitude − 30°, longitude − 90°). The hydraulic engineering required was discussed, and Lowell painted moving verbal portraits of a race of superior beings, engaged in heroic attempts to maintain their civilisation on a dying planet. Lowell's ideas were incorporated into fictional form by Edgar Rice Burroughs, in a series of books about John Carter, a terrestrial adventurer cavorting on Mars, which introduced Lowell's ideas to an even larger public.

Much more sober evidence for life on Mars was seen by some astronomers in the characteristic colour changes which follow the Martian spring. For many years it had been noticed that, as the Martian ice cap recedes, a wave of darkening flows slowly across the planet's surface from the polar regions to the equator and on into the opposite hemisphere. The wave of darkening was accompanied by a wave of water vapour formed from the melting of the ice cap, which was carried by atmospheric circulation gradually to the opposite pole. This would have provided the necessary humidity required for a Martian flora to bloom and darken the planet's surface; and the darkening was therefore likened by many to the rapid growth of vegetation in terrestrial deserts after a sudden period of rain.

During the 60s a number of spacecraft, including *Mariner II* and *IV*, were sent past Mars to photograph its surface and to carry out measurements to determine more accurately the composition of its atmosphere. These fly-by space flights put paid completely to the canal theories of Lowell and, on the whole, were rather discouraging to those who were hoping to find some signs of life on Mars. Mars proved far more arid than had been expected and there was far less oxygen in the atmosphere than had been previously believed. But the enthusiasm of the more ardent exponents of a Martian biota were not too disillusioned because none of the findings of any of the fly-by spacecraft were able to exclude completely the possibility of life. Only by placing a spacecraft capable of sampling the soil on the Martian surface could the question be finally decided.

The *Viking* mission to Mars was of truly historic significance. *Time* magazine caught the mood of mission control at the time; in the words of Gerald Soffen, one of the *Viking* project scientists:[8]

How many times does Columbus arrive in history? . . . We've just witnessed one of the arrivals. We are a privileged generation.

But *Viking* was more than any voyage of Columbus. Not only might it have decided finally the question of life on Mars but, far more significantly, because Mars is the only planet in our solar system capable of supporting any sort of life, the *Viking* mission was likely therefore to be the only chance humanity would ever have of establishing the existence of extraterrestrial life by direct contact. It was small wonder that the scientists at Pasadena broke out into spontaneous applause at its touch down on the Martian surface. For anyone interested in the nature and origin of life, the reports that were sent out from Pasadena over the next few months were of unprecedented significance.

At issue was the fundamental question as to whether life is unique to Earth. Science can only deal with repeatable or recurrent events. A unique or very improbable event can never be the subject of scientific investigation. If life is unique to Earth then this means that it has only arisen once in all cosmic history, which would essentially exclude any sort of scientific approach to the problem of its origin. Before the study of the origin of life can be put on a serious scientific footing, the possibility that life is unique to Earth has to be excluded.

If *Viking* had found evidence of life on Mars it would have put paid once and for all to the possibility of life being unique to Earth. It would have brought the question of the origin of life fully into the domain of science. A very serious philosophical shadow clouding the whole issue of the origin of life would have been removed. It would have provided powerful evidence that there is a probable route to life through transitional proto-cells or other intermediate states of matter; that life is common in the universe; and that it has arisen on other occasions by perfectly natural means wherever the planetary environment was suitable – and therefore provided massive backing if not actual proof of the whole traditional evolutionary view of life. Believers in evolution in the fall of 1976 could only wish for one message from *Viking* – the demonstration of life on Mars.

To detect life the *Viking* spacecraft was equipped with one of the technological marvels of our age: a one cubic foot box capable of carrying out almost as many experiments as a full-sized university biology lab, a masterpiece of micro-miniaturization housing some forty thousand components, pumps, chambers, filters and electronic parts. Four different experiments were carried out to detect life.

In one experiment a Martian soil sample was incubated in simulated Martian sunlight under an atmosphere containing radioactive carbon dioxide (labelled with carbon 14). After five days the soil was examined to determine whether any of the radioactive carbon dioxide had been incorporated by micro-organisms into organic compounds in the soil. In a second experiment, called the labelled-release experiment, a sample of Martian soil was moistened with a nutrient rich in vitamins and amino acids labelled with radioactive carbon 14. Any micro-organism in the Martian soil which utilized these nutrients for growth and metabolism would be likely to release radioactive waste gases into the atmosphere of the incubation chamber. Over the days following the addition of the nutrients the gases in the chamber were monitored by sensitive radioactivity detectors for any signs of carbon 14. In the third experiment, a Martian soil sample was incubated with another sort of nutrient broth and the atmosphere in the incubation chamber monitored for hydrogen methane, oxygen and carbon dioxide, gases generally produced by the proccesses of life.

The fourth experiment was perhaps the most crucial of all. This was designed to detect the existence of organic compounds in the Martian soil. A sample of soil was analysed, using gas chromatography and a mass spectrometer. So sensitive was this instrument that it was capable of detecting vanishingly small quantities of most common organic compounds, levels as little as a million millionths of a gram in some cases. The placement of a gas chromatograph and a mass spectrometer in a cubic foot box and operating it from more than two hundred million miles away was one of the technological triumphs of the *Viking* mission.

The first messages beamed from Mars gave rise to great excitement as they seemed to indicate the presence of life. The project scientists, many of them deeply committed to the traditional evolutionary explanation of life's origin, were euphoric. *Time* describes the mood in those heady few days:[9]

> It was an electrifying announcement. At a hastily called press conference at the Jet Propulsion Laboratory in Pasadena, Calif., last weekend, Viking Scientist Harold Klein reported that the newly begun biology experiments aboard the Mars lander had already shown a strange process – perhaps life – going on in the Martian soil. Said Klein: "We have at least preliminary evidence of a very active surface material. It looks at first indication very much like biological activity."
>
> Said a Viking spokesman: "If there is life on Mars, this is what it should be doing."

But the euphoria did not last. Careful analysis of the results suggested non-biological explanations to some of the *Viking* scientists. Moreover, the crucial fourth experiment to detect organic compounds in the soil entirely failed to find them even in the minutest quantities. Six weeks later, no one was so certain. The issue of *Time*, September 20, reports Klein, leader of the biology team, in a more cautious mood:[10]

> Mars is telling us something. The question is whether Mars is talking with a forked tongue or giving us the straight dope.

As the weeks passed in the fall of 1976, as the biology experiments yielded at best ambiguous results and as more and more non-biological explanations were offered to explain the results when repeated analyses of Martian soil samples failed to find any evidence of organic compounds, the realization began to dawn that, after all the hyperbole, after all that had been written and said by the enthusiasts of exobiology, Mars was going to prove a lifeless world.

A year after *Viking* first touched down on Mars, the great majority of scientists involved had become resigned to the idea that the soils of the red planet, while very unusual in many respects and exhibiting a number of interesting chemical properties, some of which had mimicked life if the first few days of the biology experiments, contained no real signs of life.

If Mars had been a planet very similar to the Earth with large quantities of surface water and a similar climate and atmosphere, then the failure to find life would have been a tremendous blow to the idea that life is an inevitable outcome of ordinary chemical processes operating on any suitable planetary surface. As it is, the present Martian climate is extremely cold and dry and only one or two of the hardiest kinds of micro-organisms could survive there. However, there is evidence for the existence of surface water in the past in structures which look like dried out river beds, so the climate may have been much warmer and wetter in former times. There is clear evidence of volcanic activity on Mars, and in the early stages of the planet's history the gases released from the Martian interior may have been similar to those which outgassed from the early Earth.

It is very dangerous to draw any firm conclusions from the limited geological data available from the *Viking* mission, and obviously constructing models of the development of the ancient Martian

258   Evolution: A Theory in Crisis

atmosphere is an even more speculative business than in the case of the Earth. Nevertheless, there is a possibility that Mars has at times experienced climatic conditions perhaps not too disimilar to those which have existed on Earth, and in this context the absence of any life forms is not so easy to account for if life really is as probable as most evolutionary theorists maintain.

The absence of life on Mars means that, as it is probably the only planet in the solar system capable of harbouring any sort of life, further planetary exploration is unlikely to establish the existence of extraterrestrial life. This leaves the detection of extraterrestrial intelligent activity from, say, radio signals as perhaps the only way to resolve the question of whether life is unique to Earth. As mentioned above, over the past fifteen years, a serious start has been made towards the detection of extraterrestrial civilizations and several nearby stars and segments of the sky have been carefully scanned for radio signals. To date all the surveys, including the well-publicized "OZMA Project", have proved negative.

But it is not only their radio signals we would expect to detect. The American theoretical physicist, Freeman Dyson, of the Institute of Advanced Study in Princeton, has raised the interesting point that if there were indeed highly developed technological civilizations extant in the universe we should expect to find other signs of their existence. It is worth quoting some of Dyson's arguments at length:[11]

> My argument begins with the following idea. If it is true, as many chemists and biologists believe, that there are millions of places in the universe where technology might develop, then we are not interested in guessing what an average technological society might look like. We have to think instead of what the most conspicuous out of a million technologies might look like. The technology which we have a chance to detect is by definition one which has grown to the greatest possible extent. So the first rule of my game is: think of the biggest possible artificial activities, within limits set only by the laws of physics and engineering, and look for those. I do not need to discuss questions of motivation, who would want to do these things or why. Why does the human species explode hydrogen bombs or send rockets to the moon? It is difficult to say exactly why. My rule is, there is nothing so big nor so crazy that one out of a million technological societies may not feel itself driven to do, provided it is physically possible.
>
> There are two more rules of my game which I shall state explicitly. Others may like to choose different rules, but I think mine are reasonable and I shall defend them if anybody objects to them.

Second rule: I assume that all engineering projects are carried out with technology which the human species of the year 1965 A.D. can understand. This assumption is totally unrealistic. I make it because I cannot sensibly discuss any technology which the human species does not yet understand. Obviously a technology which has existed for a million years will be likely to operate in ways which are quite different from our present ideas. However, I think this rule of allowing only technology which we already understand does not really weaken my argument. I am presenting an existence proof for certain technological possibilities. I describe crude and clumsy methods which would be adequate for doing various things. If there are other more elegant methods for doing the same things, my conclusions will still be generally valid.

My third rule is to ignore questions of economic cost.

Dyson goes on to argue that some civilizations, either in their quest for energy or for purposes obscure to us, would inevitably create artifacts or change their planetary systems on such a colossal scale that they would be visible across hundreds of millions of light years. Given time, there is no reason why even the energy of stars might not be utilized and the structure of whole galaxies drastically changed. But the heavens are curiously empty of any artifact-like phenomena and Dyson concludes:[12]

At the end of all these delightful speculations, we come back to the hard question, why do we not see in our galaxy any evidence of large-scale technology at work? In principle there might be two answers to this question. Either we do not see technology because none exists, or we do not see it because we have not looked hard enough. After thinking about this problem for a long time, I have come reluctantly to the conclusion that the first answer is the more probable one. I have the feeling that if an expanding technology had ever really got loose in our galaxy, the effects of it would be glaringly obvious. Starlight instead of wastefully shining all over the galaxy would be carefully damned and regulated. Stars instead of moving at random would be grouped and organized. In fact, to search for evidence of technological activity in the galaxy might be like searching for evidence of technological activity on Manhattan Island. Nothing like a complete technological takeover has occurred in our galaxy. And yet the logic of my argument convinces me that, if there were a large number of technological societies in existence, one of them would probably have carried out such a take-over.

> So in the end I am very skeptical about the existence of any extra-
> terrestrial technology. Maybe the evolution of life is much less prob-
> able event than the molecular biologists would have us believe.

Since the *Viking* mission there seems to have been a discernible waning in the confidence of exobiologists and space scientists in the possibilities of life in space. It may well be that in the future the *Viking* mission will be seen to represent a pyschological high water mark in the tide of belief in the idea of extraterrestrial life.

At present, if we are to exclude UFOs and the claims of Von Däniken and his fellow travellers, there is not a shred of evidence for extraterrestrial life, and there is no way of excluding the possibility of life being unique to Earth with all the philosophical consequences this entails. Extraterrestrial life may exist but it does not seem to be as ubiquitous as some would like to believe. Its apparent rarity and even its possible absence altogether are perfectly compatible with the possibility that there is no continuum of functional forms through which the gradual evolution of the cell might have occurred – just a yawning gulf which can only be crossed in one vastly improbable leap and as we shall see below, recent attempts to envisage how the gap might have been crossed tend increasingly to bear this out.

The basic outline of the traditional evolutionary scenario is well known. It has been expounded over and over again during the past twenty years on television, in the press, in popular scientific journals. The first stage on the road to life is presumed to have been the build-up, by purely chemical synthetic processes occurring on the surface of the early Earth, of all the basic organic compounds necessary for the formation of a living cell. These are supposed to have accumulated in the primeval oceans, creating a nutrient broth, the so-called "pre-biotic soup". In certain specialized environments these organic compounds were assembled into large macromolecules, proteins and nucleic acids. Eventually, over millions of years, combinations of these macromolecules occurred which were endowed with the property of self-reproduction. Then driven by natural selection ever more efficient and complex self-reproducing molecular systems evolved until finally the first simple cell system emerged.

The existence of a prebiotic soup is crucial to the whole scheme. Without an abiotic accumulation of the building blocks of the cell no life could ever evolve. If the traditional story is true, therefore, there must have existed for many millions of years a rich mixture of organic compounds in the ancient oceans and some of this material would very likely have been trapped in the sedimentary rocks lain down in the seas of those remote times.

Yet rocks of great antiquity have been examined over the past two decades and in none of them has any trace of abiotically produced organic compounds been found. Most notable of these rocks are the "dawn rocks" of Western Greenland, the earliest dated rocks on Earth, considered to be approaching 3,900 million years old. So ancient are these rocks that they must have been lain down not long after the formation of the oceans themselves and perhaps only three hundred to four hundred million years after the actual formation of the Earth. And the Greenland rocks are not exceptional. Sediments from many other parts of the world dated variously between 3,900 million years old and 3,500 million years old also show no sign of any abiotically formed organic compounds. As on so many occasions, paleontology has again failed to substantiate evolutionary presumptions. Considering the way the prebiotic soup is referred to in so many discussions of the origin of life as an already established reality, it comes as something of a shock to realize that there is absolutely no positive evidence for its existence.

On top of the failure to find empirical evidence of abiotically-produced organic compounds there are theoretical difficulties as well. In the presence of oxygen any organic compounds formed on the early Earth would be rapidly oxidized and degraded. For this reason many authorities have advocated an oxygen-free atmosphere for hundred of millions of years following the formation of the Earth's crust. Only such an atmosphere would protect the vital but delicate organic compounds and allow them to accumulate to form a prebiotic soup. Ominously, for believers in the traditional organic soup scenario, there is no clear geochemical evidence to exclude the possibility that oxygen was present in the Earth's atmosphere soon after the formation of its crust.[13]

But even if there was no oxygen, there are further difficulties. Without oxygen there would be no ozone layer in the upper atmosphere which today protects the Earth's surface from a lethal dose of ultraviolet radiation. In an oxygen-free scenario, the ultraviolet flux reaching the Earth's surface might be more than sufficient to break down organic compounds as quickly as they were produced. Significantly, the absence of organic compounds in the Martian soil has been widely attributed to just such a strong ultraviolet flux which today continuously bombards the planet's surface. What we have then is a sort of "Catch 22" situation. If we have oxygen we have no organic compounds, but if we don't have oxygen we have none either.

There is another twist to the problem of the ultraviolet flux. Nucleic acid molecules, which form the genetic material of all modern organisms, happen to be strong absorbers of ultraviolet light and are consequently particularly sensitive to ultraviolet-induced radiation damage and mutation. As Sagan points out, typical contemporary organisms subjected to the same intense ultraviolet flux which would have reached the Earth's surface in an oxygen-free atmosphere acquire a mean lethal dose of radiation in 0.3 seconds. Moreover, he continues:[14]

> Unacceptably high mutation rates will of course occur at much lower u.v. doses, and even if we imagine primitive organisms having much less stringent requirements on the fidelity of replication than do contemporary organisms, we must require very substantial u.v. attenuation for the early evolution of life to have occurred.

The level of ultraviolet radiation penetrating a primeval oxygen-free atmosphere would quite likely have been lethal to any proto-organism possessing a genetic apparatus remotely resembling that of modern organisms.

The oxygen–ultraviolet conundrum is only one of several such theoretical objections which can be raised against the idea of an accumulation of abiotically-produced compounds on the early Earth. In the absence of empirical evidence, the existence of additional serious theoretical objections further compounds the weakening of the traditional framework.

Studies of the earliest sedimentary rocks have also produced another difficulty which tends to mitigate against the traditional picture. Thirty years ago, the earliest signs of life on Earth were the fossils of metazoan organisms in rocks no older than seven hundred million years. As the Earth was reckoned to be in the order of several thousand millions years old, there seemed therefore to be an immense interval of time for the formation of the prebiotic soup and the gradual evolution of the cell. However, since the early 1960s, the time interval has successively shrunk as evidence of life has been discovered in increasingly older rocks. Recently, an Australian group reported the remains of a simple type of algae in rocks at least 3,500 million years old, and other rocks almost as ancient in other parts of the world have also yielded evidence of life over the past few years. The time interval available for the formation of, and evolution of, the cell from the prebiotic soup has thus dramatically shrunk from thousands to at most a few hundred million years; and, worse still, while the time interval has shrunk the earliest rocks have failed to yield any evidence of a prebiotic soup.

The existence of a prebiotic soup is an absolute prerequisite for the evolutionary emergence of life on Earth, but even if good evidence for the soup had been found the problem of the origin of life would still be far from solved. The most difficult aspect of the origin of life problem lies not in the origin of the soup but in the stages leading from the soup to the cell. Between the basic building blocks, amino acids, sugars and other simple organic compounds used in the construction of the cell, and the simplest known types of living systems there is an immense discontinuity.

The American biochemist Harold Morowitz[15] has speculated as to what might be the absolute minimum requirement for a completely self-replicating cell, deriving essential organic precursors, amino acids, sugars, etc. from its environment but autonomous in every other way in terms of current biochemistry. Such a cell would necessarily be bound by a cell membrane and the simplest feasible is probably the typical bilayered lipid membrane utilized by all existing cells. The synthesis of the fats of the cell membrane would require perhaps a minimum of five proteins. Energy would be required and some eight proteins might be needed for a very simplified form of energy metabolism. A minimum of ten proteins would be required for synthesis of the nucleotide building blocks of the DNA, and for DNA synthesis. Such a cell would also require a protein synthetic apparatus for the synthesis of its proteins. If this was along the lines of the usual ribosomal system, it would require a minimum of about eighty proteins.

Such a minimal cell containing, say, three ribosomes, 4 mRNA molecules, a full complement of enzymes, a DNA molecule 100,000 nucleotides long and a cell membrane would be about 1000Å ($1Å = 10^{-8}$ cm) in diameter. According to Morowitz:[16]

This is the smallest hypothetical cell that we can envisage within the context of current biochemical thinking. It is almost certainly a lower limit, since we have allowed no control functions, no vitamin metabolism and extremely limited intermediary metabolism. Such a cell would be very vulnerable to environmental fluctuation.

The smallest known bacterial cells, Morowitz continues, have:

> . . . an average diameter of less than 3000Å. Since the minimum hypothetical cell has a diameter of over 1000Å there is a limited gap in which to seek smaller cells.

The minimal cell described above would contain sufficient DNA to code for about one hundred average sized proteins, which is close to the observed coding potential of the smallest known bacterial cells. It may be, therefore, that the tiniest of all known bacterial cells are very close to satisfying the minimum criteria for a fully autonomous cell system capable of independent replication. The complexity of the simplest known type of cell is so great that it is impossible to accept that such an object could have been thrown together suddenly by some kind of freakish, vastly improbable, event. Such an occurrence would be indistinguishable from a miracle. An estimate of just how improbable it might be is made in Chapter Thirteen.

To explain the origin of the cell in evolutionary terms it is necessary to postulate a series of far simpler cell systems, leading gradually from a solution of organic compounds through more complex aggregates of matter to the typical cell system today. The only possible precursor to the existing cell system with its wonderfully efficient translational apparatus would be one that was less perfect. This is conceded in nearly every discussion of the origin of the cell. Discussing the evolution of the translational mechanism and the tremendous complexity of the system in all present-day cells Carl Woese argues:[17]

> It is self evident that such a hierarchy is the product of a complex evolutionary process which in turn makes it essentially certain that at some stage sufficiently early in evolution the translation mechanism was a far more rudimentary thing than at present in particular *far more prone to make translation errors.*
>
> *[emphasis added]*

Obviously, a proto-cell system would be bound to have been far more prone to making translational errors when synthesizing proteins. As Woese acknowledges, "the probability of translating any gene entirely correctly was essentially zero."[18] The hypothetical proteins produced by such an imperfect translational system have been termed "statistical proteins" by him,[19] "very crudely made proteins" by Francis Crick.[20] The trouble with "crudely made proteins" is that everything we have learned about protein structure and function over the past thirty years implies that the function of a protein depends on it being very accurately manufactured and possessing exact highly specific configurations. We seem forced to have to contradict one of the basic axioms of modern biochemisty in envisaging the origin of the cell.

The difficulty of visualizing how a cell containing a primitive error-prone translational system and capable of manufacturing only "statistical proteins" could ever successfully produce a functional enzyme and be viable is met in every discussion of the origin of the translational system, as Woese admits:[21]

> How could such a cell contain any enzymes and so be viable? We cannot answer this definitely at the present state of our knowledge.

However, if evolution is to be believed, such remarkable entities simply had to exist, as he continues:[22]

> Nevertheless it is essentially a certainty that at an early enough stage in evolution such cells as these did exist and some had to be viable.

Just how the translational system could ever function utilizing "crudely made proteins" is virtually impossible to envisage, indeed it is exceedingly difficult to understand how translation in any meaningful sense could occur.

The protein synthetic system of all modern cells requires the integrated activities of nearly one hundred different proteins, all carrying out different, very specific steps in the assembly of a new protein molecule. If only a small proportion of these were "crudely made" or "statistical" it is practically impossible to accept that any protein would ever be manufactured, let alone one with a specific molecular configuration capable of performing a specific function in the cell.

It is precisely because the translation system is critically dependent on accurately made proteins that an imperfect protein synthetic system is so difficult to envisage. The same central difficulty is met by Emile Zuckerkandl:[23]

> The evolution of an efficient apparatus of translation strongly suggests that the "discovery" of efficient enzymic molecules preceded the last phases of construction of this apparatus.

Just how efficient enzymes could have been manufactured before an efficient translational system was in existence is absolutely mystifying, as he confesses:[24]

> The sophisticated contemporary type of translational apparatus could hardly have properly functioned on the basis of proteins not yet endowed with the property of specificity of interaction.

So efficient enzymes must have preceded an accurate translational system but efficient enzymes are absolutely dependant on an accurate translational system! Zuckerkandl is forced to conclude:[25]

> We seem to be forced to think that the evolution of functional efficiency in proteins and the setting up of the modern translational apparatus proceeded step by step and went hand in hand. Each progress in functional efficiency in proteins caused progress in fidelity of translation, and vice versa.

So Zuckerkandl, like Woese, is forced to imagine a hypothetical cell in which efficient proteins evolved in a system basically incapable of their manufacture. In Woese's words: "The primitive cell was faced with the seeming paradox that in order to develop a more accurate translational apparatus it had first to translate more accurately."[26]

If translation is inaccurate, this leads in turn to a more inaccurate translational apparatus which leads inevitably to further inaccuracies, and so forth. Each imperfect cycle introduces further errors.[27] To improve itself, such a system would have to overcome its fundamental tendency to accumulate errors in exponential fashion. The very cyclical nature of cellular replication guarantees that imperfections inexorably lead to autodestruction. It is difficult enough to see how an imperfect translational system could ever have existed and achieved the synthesis of one single protein let alone the many necessary for the life of the cell. That such a cell might undergo further evolution, improving itself by "selecting" advantageous changes which would be inevitably lost in the next cycle of replication, seems contradictory in the extreme.

Modern organisms get by despite mutations because the rate of mutation is low. The rate in all organisms from bacteria to mammals has been estimated for various loci at between 10^{-9} to 10^{-10} per base pair copied when DNA is replicated.[28] This is low and infrequent enough to guarantee that all the vital machinery concerned with the organism's self-duplication, including its protein synthetic apparatus, can be duplicated perfectly as the entire system is specified in less than 10^6 base pairs. Moreover, complex organisms today often exhibit redundancy in the genes specifying for their essential protein synthetic machinery.

However, if the mutation rate is raised by, say, irradiation, then certainly this leads to an accumulation of errors down a chain of replication which is ultimately lethal to the clone of irradiated individuals (this can be experimentally demonstrated with microorganisms). When the mutation rate is very high, no living system can avoid the path to autodestruction. Each cycle increases the "noise" and erases crucial information, like a series of increasingly poor photocopies; ultimately, the text becomes illegible. That an error-prone translational system would lead inevitably to self-destruction is not only a theoretical prediction but also a well-established empirical observation.

The only possible escape from the paradox of self-destruction is to envisage a very high level of redundancy in the proto-cell. Each proto-cell would require sufficient copies of each gene to ensure that, despite inefficient gene copying and inefficient gene translation, it would always manage to pass on some correct genes to the next generation, and to manufacture at least enough correct proteins to carry out its vital functions.

One of the implications of such redundancy would inevitably be the existence of vast quantities of randomly coiled junk protein in the cell, a reality which is almost certainly incompatible with cell function. Such junk proteins would stick randomly to various entities in the cell and tend to combine in a haphazard manner with all its essential molecular machinery, choking it in a chaos of tangled fibres. As is seen in sickle cell anaemia, even in a modern cell like the human red blood cell, the change of one amino acid in one protein can cause massive distortion of the entire cytoarchitecture of the cell. Moreover, the energy burden posed by massive redundancy would be immense.

In envisaging the halfway world to life we have one of the most dramatic examples in all of nature of one of Cuvier's "incoherent combinations" – a contradictory entity which could never have functioned in the real world. Just as a bird feathered with the frayed scales of pro-avis would plummet to the ground, so a cell burdened with inefficient proteins, an error-prone code, and choked with junk would grind instantly to a halt.

In considering the origin of the translational system, evolution theory seems to have reached a sort of nemesis, for the problem is to all intents and purposes insoluble in terms of modern biochemical knowledge. That the profundity of the problem of the origin of translational systems has stretched the evolutionary framework to breaking point is conceded by Monod:[29]

> The development of the metabolic system which as the primordial soup thinned must have "learned" to mobilize chemical potential and to synthesise the cellular components poses Herculean problems. So does the emergence of a selectively permeable membrane without which there can be no viable cell. But the major problem is the origin of the genetic code and of its translational mechanism. Indeed it is not so much a problem as a veritable enigma.

Another Nobel Prize winner, biochemist Francis Crick, in his recent book, *Life Itself*, concedes:[30]

> An honest man, armed with all the knowledge available to us now, could only state that in some sense, the origin of life appears at the moment to be almost a miracle, so many are the conditions which would have had to have been satisfied to get it going.

The origin of life is actually far more difficult to envisage than the above discussion implies. There is much more to the cell than the "mere" origin of the protein synthetic apparatus. In fact, the protein synthetic mechanism cannot function in isolation but only in conjunction with other complex subsystems of the cell.

Without a cell membrane the components of the protein synthetic apparatus could not be held together. The integrity of the cell membrane, however, depends on the existence of a protein synthetic apparatus capable of synthesizing the protein components of the membrane and the enzymes required for the synthesis of its fat components. However, the protein synthetic apparatus consists of a number of different components and can only function if these are held together by a membrane: two seemingly unbreakable inter-dependent systems. To continue, the protein synthetic apparatus also requires energy. The provision of energy depends on the coherent activity of a number of specific proteins capable of synthesizing the energy-rich phosphate compounds – proteins which are themselves manufactured by the protein synthetic apparatus. A further couple of interdependent cycles! As we have seen, the information for the specification of all the protein components of the cell, including those of the protein synthetic apparatus, is stored in the DNA. However, the extraction of this information is dependent on the proteins of the protein synthetic apparatus – yet again another set of interdependent cycles.

It is not only biochemists who have difficulty in envisaging the design of a simple self-replicating system. Eminent engineers and mathematicians, such as von Neumann,[31] who have considered theoretically the general abstract design of self-replicating automata have shown that any automaton sufficiently complex to reproduce itself would necessarily possess certain component systems which are strictly analogous to those found in a cell. One component would be an automatic factory capable of collecting raw materials and process-ing them into an output specified by a written instruction. This is the analogue of the ribosome. Another component would be a duplicator, an automaton which takes the written instruction and copies it. This is the analogue of the DNA replicating system. Another component would be a written instruction containing the specification for the complete system, which is the analogue of the DNA (see Figure over-leaf).

The fact that artificial automata and living organisms both have to conform to the same general design to meet the criteria for self-replication tends to reinforce the feeling that perhaps no system simpler than the cell system can exist which can undergo genuine autonomous self-duplication.

The difficulty that is met in envisaging how the cell system could have originated gradually is essentially the same as that which is met in attempting to provide gradual evolutionary explanations of all the other complex adaptations in nature. It is perfectly obvious, in the case of the feather, that function as an aerofoil is impossible unless

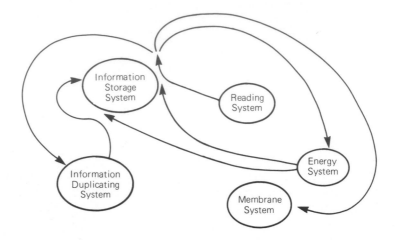

the hooks fit the barbules, that is, unless the components are exquisitely coadapted to function together. It is the same in the case of the avian lung or in the case of the wing of a bat, and it is the same in the case of human artifacts such as a watch which can only function when all the cogwheels fit together, and it is the same in the case of sentences. The problem of the origin of life is not unique – it only represents the most dramatic example of the universal principle that complex systems cannot be approached gradually through functional intermediates because of the necessity of perfect coadaptation of their components as a pre-condition of function. Transitions to function are of necessity abrupt. The origin of life problem lends further support to the notion that the divisions of nature arise out of the necessities rooted in the logic of the design of complex systems.

The only alternative is to consider the possibility of saltation. However, the probability of a sudden fortuitous event assembling the first cell *de novo* has generally struck most biologists as outrageously improbable. Yet, if gradualism is impossible, there may be no alternative but to presume that such an extraordinarily lucky accident was responsible for creating the first cell. Monod has also raised the possibility:[32]

> Life appeared on earth: what, before the event, were the chances that this would occur? The present structure of the biosphere certainly does not exclude the possibility that . . . its a priori probability was virtually zero.

Crick[33] has also recently conceded that life may after all be very improbable and has turned to an interesting variation on the saltational alternative, the idea that life was orginally seeded on earth from space – the idea of panspermia. The British astronomer Fred Hoyle and his colleague Chandra Wickramasinghe have, for similar reasons, also recently raised the possibility of panspermia. Nothing illustrates more clearly just how intractable a problem the origin of life has become than the fact that world authorities can seriously toy with the idea of panspermia.[34]

The failure to give a plausible evolutionary explanation for the origin of life casts a number of shadows over the whole field of evolutionary speculation. It represents yet another case of a discontinuity where a lack of empirical evidence of intermediates coincides with great difficulty in providing a plausible hypothetical sequence of transitional forms. It therefore tends to reinforce the possibility that the discontinuities of nature may be much more fundamental than merely the artefactual result of random sampling that evolution implies.

Moreover, the seemingly intractable difficulty of explaining how a living system could have gradually arisen as a result of known chemical and physical processess raises the obvious possibility that factors as yet undefined by science may have played some role. Such a concession is, of course, the thin end of a very dangerous wedge for once it is conceded that *one* evolutionary event has involved novel and unknown processes and has been more than a matter of chance and selection then the whole framework of Darwinian evolution is threatened. An obvious extrapolation is suggested – may not these unidentified processes have been involved in other problematical areas of evolution?

On the whole, the new biochemical picture has not had the effect that evolutionary theorists might have hoped. It has not blurred the distinction between living and non-living objects. The recently revealed world of molecular machinery, of coding systems, of informational molecules, of catalytic devices and feedback control, is in its design and complexity quite unique to living systems and without parallel in non-living nature.

NOTES

1. Monod, J. (1972) *Chance and Necessity*, Collins, London, p 134
2. Huxley, T. H. (1970) "Biogenesis and Abiogenesis" in *Collected Essays of T. H. Huxley* (1894) 9 vols, Macmillan and Co, London, vol 8, pp 229–71, see p 256.
3. Sagan, C. (1977) *Intelligent Life in the Universe*, Picador by Pan Books Ltd, London, p 358.
4. Huygens, C. (1670) "New Conjectures Concerning the Planetary Worlds, Their Inhabitants and Productions", cited by Sagan in *Intelligent Life in the Universe*, p 214.
5. ibid, p 356.
6. Whewell, W. (1854) *The Plurality of Worlds*, John W. Parker and Sons, London, pp 24 and 286.
7. Sagan, op cit, p 276.
8. Soffen, G. (1976), cited in "Mars: The Riddle of the Red Planet", *Time Magazine*, 2 August, p 16.
9. *Time Magazine* (1976) "Viking: The First Signs of Life", 9 August, p 48.
10. Klein, H. (1976) "Looking for the Bodies", *Time Magazine*, 20 September, p 74.
11. Dyson, J. F. (1966) "The Search for Extraterrestrial Technology" in *Perspectives in Modern Physics*, ed R. E. Marshak, John Wiley and Sons, Chichester and New York, pp 643–44.
12. ibid, pp 653–54.
13. Dimroth, E. and Kimberley, M. M. (1976) "Pre-Cambrian Atmospheric Oxygen: Evidence in the Sedimentary Distribution of Carbon, Sulfer, Uranium and Iron", *Canadian Journal of Earth Sciences*, 13: 1161–85, see p 1161.
14. Sagan, C. (1973) " Ultraviolet Selection Pressure on the Earliest Organisms", *J. Theoretical Biology*, 39: 195–200, see p 197.
15. Morowitz, H. J. (1966) "The Minimum Size of Cells" in *Principles of Bio-molecular Organisation*, eds G. E. W. Wostenholme and M. O'Connor, J. A. Churchill, London, pp 446–59.
16. ibid, p 456.
17. Woese, C. (1965) "On the Origin of the Genetic Code", *Proc. Natl. Acad. Sci. U.S.*, 54: 1546–52, see p 1548.
18. ibid, pp 1548–49.
19. ibid, p 1549.
20. Crick, F. H. C. (1968) "The Origin of the Genetic Code", *J. Mol. Biol.*, 38: 367–79, see p 357.
21. Woese, op cit, p 1549.
22. ibid.
23. Zuckerkandl, E. (1975) "The Appearance of New Structures and Functions in Proteins During Evolution", *J. Mol. Evolution*, 7: 1–57, see p 30.
24. ibid, p 31.
25. ibid.
26. Woese, op cit, p 1549.
27. ibid.
28. Drake, J. W. (1973) *The Molecular Basis of Mutation*, Holden-Day Inc, San Francisco, see Table 5.3.

29. Monod, op cit, p 135.
30. Crick, F. (1981) *Life Itself*, Simon and Schuster, New York, p 88.
31. Von Neumann, J. (1966) *Theory of Self-Reproducing Automata*, University of Illinois Press, Urbana.
32. Monod, op cit, p 136.
33. Crick, F. and Orgel, L. E. (1973) "Directed Panspermia", *Icarus*, 19: 341–46.
34. Hoyle, F. and Wickramasinghe, C. (1981) *Evolution from Space*, J. M. Dent and Sons, London.

CHAPTER 12
A Biochemical Echo
of Typology

... as we here and there see a thin straggling branch springing from a fork low down in a tree and which by some chance has been favoured and is still alive on its summit so we occasionally see an animal like Lungfish which in some small degree connects by its affinities two large branches of life.

We have seen that at a morphological level the pattern of nature seems to correspond reasonably well with the old nineteenth-century typological model. Nearly all known groups appear to be isolated and well defined and clear sequential patterns whereby one class is linked to another through linear series of transitional forms are virtually unknown. Moreover, classification procedures invariably result in orderly hierarchic schemes from which overlapping classes indicative of sequential relationships are emphatically absent.

However, no matter how much the diversity of nature may appear to conform to the theory of types at a morphological level, no matter how much all cats, all birds, all angiosperms, all mammals or all vertebrates may seem to be equally representative of their respective groups, there is no way of quantifying such conclusions. Judging relationships in terms of morphological characteristics is bound to involve an element of subjectivity. On purely morphological grounds there is no way of measuring the *exact* distance between two organisms in strictly mathematical terms. We cannot, for example, quantify the difference between a cat and a dog and compare it with, say, the difference between a cat and a mouse. We assume that a cat and a dog are closer than a cat and a mouse, but how secure are such judgments?

There is also simply no way of making a quantitative measurement of complexity at a morphological level. A mammal may "look" more complex than a fly, but whether this is true and, if it is, how much more complex in strictly quantitative terms cannot be determined on grounds of morphology. Again, the vertebrate central nervous system gives every "appearance" of being a vastly complex system, but for all we know it may require less genetic information to specify for the vertebrate brain than for the pentadactyl limb!

From the founding of modern biology by Linnaeus in the eighteenth century right up to the 1960s, the only way biologists had of classifying organisms and assessing the differences between species was by comparing their structure at a gross morphological level. Comparative biology was no more nor less than comparative anatomy.

The molecular biological revolution has dramatically changed this situation by providing an entirely new way of comparing organisms at a biochemical level. In the late 1950s it was found that the sequence of a particular protein, such as, say, haemoglobin, was not fixed but varied considerably from species to species. The amino acid sequence of a protein from two different organisms can be readily compared by aligning the two sequences and counting the number of positions where the chains differ. In exactly the same way two sequences of letters can be compared. For example, sequences A and B in the diagram below differ at four positions.

(A) C D K N I A A T Y L V G H I T T E N B Y
(B) C B K N I D A T Y L V G H I C T E M B Y
     1   2   3   4

There are twenty letters in each of the two sequences above so they can be said to exhibit a twenty percent sequential divergence.

Similarly, the differences between two proteins can be quantified exactly and the results of these measurements can provide an entirely novel approach to measuring the differences between species. The list below gives part of the sequence of the protein cytochrome C from a variety of species:

Horse	Gly-Leu-Phe-Gly-Arg-Lys-Thu-Gly-GluNH$_2$-Ala-Pro
Pig or cow	Gly-Leu-Phe-Gly-Arg-Lys-Thr-Gly-GluNH$_2$-Ala-Pro
Kangaroo	Gly-Ile-Phe-Gly-Arg-Lys-Thr-Gly-GluNH$_2$-Ala-Pro
Human	Gly-Leu-Phe-Gly-Arg-Lys-Thr-Gly-GluNH$_2$-Ala-Pro
Chicken	Gly-Leu-Phe-Gly-Arg-Lys-Thr-Gly-GluNH$_2$-Ala-Glu
Tuna	Gly-Leu-Phe-Gly-Arg-Lys-Thr-Gly-GluNH$_2$-Ala-Glu
Moth	Gly-Phe-Gly-Arg-His-Thr-Gly-GluNH$_2$-Ala-Pro-Gly-Phe-Tyr
Yeast	Gly-Ile-Phe-Gly-Arg-His-Ser-Gly-GluNH$_2$-Ala-GluNH$_2$

As work continued in this field, it became clear that each particular protein had a slightly different sequence in different species and that closely related species had closely related sequences. When the haemoglobin sequences in different mammals, such as man and dog, were compared the sequential divergence was about twenty percent, while, when the haemoglobin in two dissimilar species such as man carp were compared, the sequential divergence was found to be about fifty percent.

It was also found that different types of proteins exhibited different degrees of interspecies variation. Cytochrome C, for example, varied less between species than haemoglobin. While the haemoglobin sequences of man and dog differed by twenty percent, their cytochrome sequences varied by only five percent, and while the haemoglobin sequences of man and carp varied by fifty percent, their cytochrome sequences varied by only thirteen percent. Yet whichever protein was chosen, organisms that were close in terms of their haemoglobin sequences were also close in terms of their cytochromes, and the same was true of all other proteins examined.

These results showed that not only did organisms vary at a morphological level in terms of their gross anatomy, but that they also varied at a molecular level as well. It became increasingly apparent as more and more sequences accumulated that the differences between organisms at a molecular level corresponded to a large extent with their differences at a morphological level; and that all the classes traditionally identified by morphological critera could also be detected by comparing their protein sequences. Among the vertebrates, for example, all the major classes identified by morphological criteria can also be readily identified on the basis of molecular comparisons.

Armed with this new technique, biology at last possessed a strictly quantitative means of measuring the distance between two species and of determining the patterns of biological relationships. If it is true, as typology implied, that all the members of one type, however superficially divergent, always conform exactly to the basic eidos of their type, all possessing equally and in full measure all the defining character traits of their type and all standing therefore equidistant in all important aspects of their biological design from the members of other types, might this principle of equidistance be revealed by these new molecular studies? If the divisions in the nature were really as orderly as early nineteenth-century biologists insisted, might this overall orderliness be confirmed by the new field of comparative biochemistry?

On the other hand, the new molecular approach to biological relationships could potentially have provided very strong, if not irrefutable, evidence supporting evolutionary claims. Armed with this new technique, all that was necessary to demonstrate an evolutionary relationship was to examine the proteins in the species concerned and show that the sequences could be arranged into an evolutionary series. In the diagram below it is obvious that sequence B is intermediate between sequences A and C.

A	*B*	*C*
A	A	A
B	B	C
T	T	T
W	W	W
V	S	S
Y	Y	Y
H	H	H
K	L	L
D	P	P
E	E	T

It is possible to arrange the letter strings in a series where B is intermediate between A and C and to postulate either that A evolved into B and B into C or that B evolved into A and C or C into B into A.

$$A \rightarrow B \rightarrow C \qquad \begin{matrix} A & & C \\ \nwarrow & & \nearrow \\ & B & \end{matrix} \qquad C \rightarrow B \rightarrow A$$

Whichever theory is correct, such sequential arrangements suggest evolutionary relationships.

The prospect of finding sequences in nature by this technique was, therefore, of great potential interest. Where the fossils had failed and morphological considerations were at best only ambiguous, perhaps this new field of comparative biochemistry might at last provide objective evidence of sequence and of the connecting links which had been so long sought by evolutionary biologists.

However, as more protein sequences began to accumulate during the 1960s, it became increasingly apparent that the molecules were not going to provide any evidence of sequential arrangements in nature, but were rather going to reaffirm the traditional view that the system of nature conforms fundamentally to a highly ordered hierarchic scheme from which all direct evidence for evolution is emphatically absent. Moreover, the divisions turned out to be more mathematically perfect than even most die-hard typologists would have predicted.

To understand the subsequent pattern that has been revealed by these comparative studies we might start by reviewing the evidence provided by the protein cytochrome C, one of the proteins intimately connected with the production of cellular energy. Because of its fundamental role in biological oxidation it occurs in a wide range of organisms ranging from bacteria to mammals.

All cytochrome C molecules are about one hundred amino acids long, have the same 3D conformation and possess an identical active site or hydrophobic pocket specifically designed to complex tightly with the small iron-containing organic compound haem. Yet, despite the profound correspondence in the basic design of all cytochrome C molecules, their amino acid sequences vary in different organisms. The amino acid sequences of cytochrome C have now been determined in a wide variety of organisms including bacteria, fungi, higher plants and vertebrates.

When comparing a considerable number of sequences, it is convenient to present the data in the form of a percent sequence difference matrix. In the *Dayhoff Atlas of Protein Structure and Function* (1972 edition) there is a matrix with nearly 1089 entries showing the percent sequence difference between thirty-three different cytochromes taken from very diverse species. Part of this matrix is shown in Figure 12.1.

Examination of the percent sequence difference matrix reveals that it is possible to use the cytochrome sequences to classify species into groups and that these groups correspond precisely to the groups arrived at on traditional morphological grounds. It is also apparent that the sequential divergence becomes greater as the taxonomic distance between organisms increases, a finding that would again have been predicted from traditional taxonomic considerations. For example, between horse and dog (two mammals) the divergence is six percent, between horse and turtle (two vertebrates) the divergence is eleven percent, and between horse and fruit fly (two animals) the divergence is twenty-two percent.

However, the most striking feature of the matrix is that each identifiable subclass of sequences is isolated and distinct. Every

	Horse	Dog	Kangaroo	Penguin	Pekin Duck	Pigeon	Turtle	Tuna	Bonito	Carp	Lamprey	Screw-worm	Silkworm	Horn Worm	Castor	Sunflower	Wheat	C. krusei	D. kloeckeri	Yeast	R. rubrum C₂
Horse	0	6	7	12	10	11	11	18	17	13	15	20	27	26	40	41	41	46	40	42	64
Dog	6	0	7	10	8	9	9	17	16	11	13	19	23	23	38	39	39	45	38	41	65
Kangaroo	7	7	0	10	10	11	11	17	17	13	16	22	26	26	38	39	42	46	41	42	66
Penguin	12	10	10	0	3	4	8	17	17	14	18	22	25	25	40	41	41	45	40	40	64
Pekin Duck	10	8	10	3	0	3	7	16	16	13	17	20	25	25	38	39	41	45	40	41	64
Pigeon	11	9	11	4	3	0	8	17	17	14	18	21	25	24	38	39	41	45	40	41	64
Snapping Turtle	11	9	11	8	7	8	0	17	16	13	18	22	26	27	38	39	41	47	42	44	64
Tuna Fish	18	17	17	17	16	17	17	0	2	8	18	22	30	28	42	43	44	43	42	43	65
Bonito	17	16	17	17	16	17	16	2	0	7	18	23	31	29	41	41	42	42	41	41	64
Carp	13	11	13	14	13	14	13	8	7	0	12	20	25	24	41	41	42	45	39	42	64
Lamprey	15	13	16	18	17	18	18	18	18	12	0	26	30	31	45	44	46	50	43	45	66
Screw-worm Fly	20	19	22	22	20	21	22	22	23	20	26	0	13	11	40	40	40	43	39	44	64
Silkworm Moth	27	23	26	25	25	25	26	30	31	25	30	13	0	5	40	40	40	43	39	44	65
Tobacco Horn Worm Moth	26	23	26	25	25	24	27	28	29	24	31	11	5	0	39	40	38	42	39	42	64
Castor	40	38	38	40	38	38	38	42	41	41	45	40	40	39	0	10	12	45	43	42	66
Sunflower	41	39	39	41	39	39	39	43	41	41	44	40	40	40	10	0	13	47	44	43	67
Wheat	41	39	42	41	41	41	41	44	42	42	46	40	40	38	12	13	0	45	41	42	66
Candida krusei	46	45	46	45	45	45	47	43	42	45	50	43	43	42	45	47	45	0	23	25	72
Debaryomyces kloeckeri	40	38	41	40	40	40	42	42	41	39	43	39	39	39	43	44	41	23	0	27	67
Baker's Yeast	42	41	42	40	41	41	44	43	41	42	45	44	44	42	42	43	42	25	27	0	69
Rhodospirillum rubrum C₂	64	65	66	64	64	64	64	65	64	64	66	66	65	64	66	67	66	72	67	69	0

Groupings: MAMMALS, BIRDS, REPTILES (TELEOSTS, CYCLOSTOMES); INSECTS; PLANTS; YEASTS; BACTERIA.

Figure 12.1: The Cytochromes Percent Sequence Difference Matrix. (from Dayhoff)[1]

MAMMALS		BIRDS		TELEOSTS	
Human	65	Chicken	64	Tuna	65
Monkey	64	Penguin	64	Bonito	64
Pig	64	Duck	64	Carp	64
Horse	64	Pigeon	64	ELASMOBRANCHS	
Dog	65	REPTILES		Dogfish	65
Whale	65	Turtle	64	CYCLOSTOMES	
Rabbit	64	Rattlesnake	66	Lamprey	66
Kangaroo	66	AMPHIBIANS			
		Bullfrog	65		

INSECTS		ANGIOSPERMS		YEASTS	
Fruit Fly	65	Nung-bean	66	Candida krusei	72
Screw-worm	64	Sesame	65	Debaryomyces kloeckeri	67
Silkworm	65	Castor	69	Baker's yeast	69
Tobacco Horn Worm Moth	64	Sunflower	69	Neurospora crassa	69
		Wheat	66		

Figure 12.2: The Molecular Equidistance of all Eucaryotic Organisms from Bacteria. *Percent sequence divergence between the cytochrome C_2 of Rhodospirillum rubrum and various eucaryotic cytochromes.* (from Dayhoff)[2]

sequence can be unambiguously assigned to a particular subclass. No sequence or group of sequences can be designated as intermediate with respect to other groups. All the sequences of each subclass are equally isolated from the members of another group. Transitional or intermediate classes are completely absent from the matrix.

A table which illustrates the dramatic absence of intermediates is seen in Figure 12.2 which lists thirty-three comparisons between the bacterial cytochrome C of *Rhodospirillum rubrum* and cytochromes of a wide variety of eucaryotic organisms. (Eucaryotic refers to organisms whose cells possess a nucleus, ie, all non-bacterial groups.) These comparisons indicate that all the eucaryotic cytochrome sequences are almost exactly the same distance from their bacterial homologue.

In the list in Figure 12.2, if three yeasts are excluded from the list, then the remaining eucaryotic cytochromes, from organisms as diverse as man, lamprey, fruit fly, wheat and yeast, all exhibit a sequence divergence of between sixty-four percent and sixty-seven percent from this particular bacterial cytochrome. Considering the enormous variation of eucaryotic species from unicellular organisms like yeasts

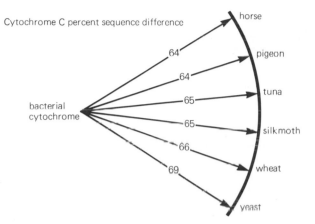

Cytochrome C percent sequence difference

bacterial
cytochrome

64

64

65

65

66

69

horse

pigeon

tuna

silkmoth

wheat

yeast

to multicellular organisms such as mammals, and considering that euraryotic cytochromes vary among themselves by up to about forty-five percent, this must be considered one of the most astonishing findings of modern science.

It means that no eucaryotic cytochrome is intermediate between the bacterial cytochrome and other eucaryotic cytochromes. As far as the bacterium is concerned, all the eucaryotes are equally distant. All the eucaryotic cytochromes are as a class isolated and unique. No intermediate type of cytochrome exists to bridge the discontinuity which divides the living kingdom into these two fundamental types. The bacterial kingdom has no neighbour in any of the fantastically diverse eucaryotic types. The "missing links" are well and truly missing.

But even among the eucaryotic cytochromes at a slightly lower taxonomic level the same isolation and uniqueness of subclasses, the same lack of intermediates are observed. Examination of the sequential divergence among eucaryotic cytochromes reveals three basic sub-groups: the yeasts, the plants, and the animals (see matrix in Figure 12.1). Each type is quite isolated. Just as there are no intermediates to bridge the gap between procaryotes and eucaryotes so there are no intermediate types among these three basic eucaryotic groups. Al-though the distance among the three eucaryotic types is less than that between the procaryotes and the eucaryotes, the divisions among the three fundamental types are no less clear and unambiguous. Each type is just as unique and isolated from the others. The yeast cyto-chromes are uniformly isolated from the cytochromes of all other eucaryotes. The same ordered isolation is seen when the plants are compared with other eucaryotes. Similarly, no animal cytochrome is intermediate between the animals and the other two eucaryotic groups.

At a still lower taxonomic level the same phenomenon is observed. From the matrix in Figure 12.1 it is clear that the insects and vertebrates are closely related, but when comparisons are made between insect species and a variety of vertebrate groups, no vertebrate group is primitive or in any sense a link between phylum Arthropoda and phylum Vertebrata. All the many diverse vertebrate types, including cyclostomes and mammals, are uniformly distant from the insects.

The diagram below gives the percentage sequence divergence between the silk moth and various vertebrate cytochromes:[3]

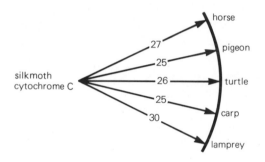

From the sequential divergence of their cytochromes it is possible to classify the living kingdom into various divisions. The primary division is clearly between bacteria and eucaryotes. The eucaryotes are subdivided into three distinct classes, yeasts, plants and animals; the animals can be subdivided into two further subclasses, insects and vertebrates.

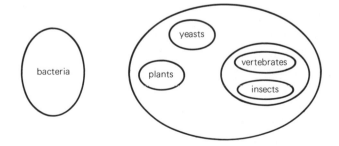

Each class is isolated and unique. No classes are intermediate or partially inclusive of other classes. The isolation of each class becomes greater as the taxonomic hierarchy is ascended, but even relatively closely related classes such as insects and vertebrates are still clearly distinguished.

The pattern of nature implied by these findings is depicted in the diagram below. In terms of their cytochromes, the three major eucaryotic kingdoms may be thought of as equidistant from a common hypothetical archetype, while within each group all the members are similarly equidistant from the hypothetical archetype of their group.

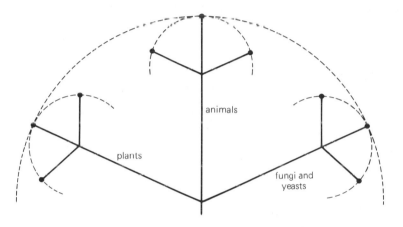

Note how closely the cytochrome pattern seems to correspond to the circumferential model of nature of the early nineteenth-century typologists.

Consider next the divisions within the vertebrate phylum. Based on the degree of similarity of their proteins, the vertebrates can be clearly divided into two fundamental divisions, the jawless cyclostomes and the higher jawed vertebrates – the fish, amphibia, reptiles and mammals.

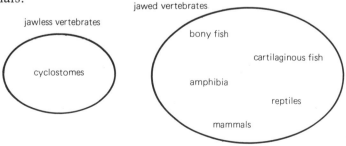

In itself, the existence of this fundamental division is not surprising, as it corresponds exactly with a traditional division based on morphological characteristics. But the strange thing about the division is the fact that although the proteins in the higher jawed vertebrate groups – fish, amphibia, reptiles and mammals – are widely divergent when they are compared with those of cyclostomes, invariably the degree of difference is always the same. The almost mathematical perfection of the isolation of the two fundamental classes at a molecular level is astonishing! The figure below gives the percent sequence difference between the haemoglobin of the lamprey and a variety of jawed vertebrates, taken from a sequence difference matrix of the vertebrate globins in the *Dayhoff Atlas of Protein Structure and Function:*[4]

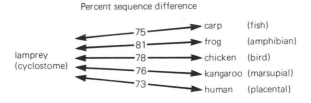

Percent sequence difference

lamprey (cyclostome)	75	carp	(fish)
	81	frog	(amphibian)
	78	chicken	(bird)
	76	kangaroo	(marsupial)
	73	human	(placental)

There is not a trace at a molecular level of the traditional evolutionary series: cyclostome → fish → amphibian → reptile → mammal. Incredibly, man is as close to lamprey as are fish! None of the higher jawed vertebrate groups is an any sense intermediate between the jawless vertebrates and other jawed vertebrate groups.

The higher vertebrate groups can also be divided into subgroups based on their degree of molecular similarity. Several basic subdivisions are revealed. One contains the many types of bony fish, another

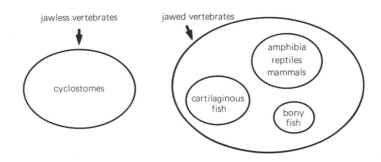

jawless vertebrates   jawed vertebrates

cyclostomes   cartilaginous fish   amphibia reptiles mammals   bony fish

the cartilaginous fish and another the terrestrial vertebrate groups, amphibia, reptiles, and mammals.

These subdivisions are not very surprising as they correspond exactly with traditional subdivisions derived from morphological studies. But again there is the same strangely ordered aspect to the pattern of the molecular divisions. When the various terrestrial vertebrate groups, amphibia, reptile, or mammal, are compared with fishes all are equally isolated. The figure below gives the percent sequence difference between cyclostome C in carp and various terrestrial vertebrates.

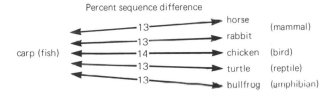

Percent sequence difference

Again, an extraordinary mathematical exactness in the degree of isolation is apparent. So, although cytochrome C sequences varied among the different terrestrial vertebrates, all of them are equidistant from those of fish. At a molecular level there is no trace of the evolutionary transition from fish → amphibian → reptile → mammal. So amphibia, always traditionally considered intermediate between fish and the other terrestrial vertebrates, are in molecular terms as far from fish as any group of reptiles or mammals! To those well acquainted with the traditional picture of vertebrate evolution the result is truly astonishing.

The terrestrial vertebrates can themselves be divided into two basic classes, by virtue of their molecular similarities. One class contains the amphibia, the other the reptiles and mammals. Again the subdivision corresponds to that based on classical morphological grounds, but whichever species are taken for comparative purposes the distance between amphibian species on the one hand and mammalian and reptilian species on the other is always the same. No amphibian species is midway between other amphibia and the reptiles and the mammals. Similarly no reptilian or mammalian species is closer to amphibia than any of the others. The vertebrate classification scheme could be represented thus:

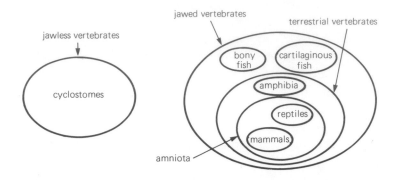

The classification system that is derived from these comparative molecular studies is a highly ordered non-overlapping system composed entirely of groups within groups, of classes which are inclusive or exclusive of other classes. There is a total absence of partially inclusive or intermediate classes, and therefore none of the groups traditionally cited by evolutionary biologists as intermediate gives even the slightest hint of a supposedly transitional character.

The molecules give no support to the traditional view of the vertebrates as a series of increasingly advanced classes leading from the cyclostomes to the mammals. In fact, when the vertebrates are compared with non-vertebrate organisms, all types are equidistant apart. The diagram below gives the percent sequence divergence between the haemoglobin in a snail and that of various vertebrate species.

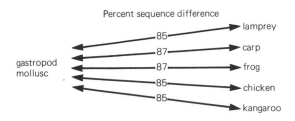

On the evidence of the protein sequences we cannot classify the lamprey as primitive with respect to other vertebrates, nor in any sense as intermediate between higher vertebrates and the invertebrates. All we can safely infer about the cyclostomes is that they represent a highly specialized and isolated vertebrate group.

Furthermore, it is not only the major divisions which can be subdivided into non-overlapping classes; the same phenomenon holds to quite minor subdivisions of the animal kingdom, even where the actual biochemical differences between species is relatively trivial.[5] For example, when classifying the primates (the monkeys, apes and man) by comparing the differences in their protein sequences,[6] the result is again an entirely non-overlapping system of classes.

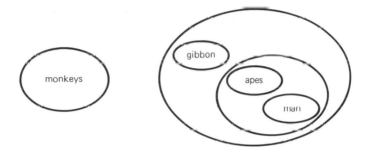

The same picture emerges when DNA or RNA sequences in different species are compared. A fascinating case which illustrates this is the recent work carried out by Carl Woese's group at the University of Illinois, which led to the discovery of a new primary kingdom consisting of an interesting group of anaerobic bacteria possessing the unique capacity to generate the gas methane.[7]

The discovery that the methanogens belonged to a new, quite distinct, living kingdom was made as Woese's group was comparing the RNA sequences of the large RNA molecules which form part of the protein synthetic apparatus (termed ribosomal RNA) in a wide variety of organisms. They showed that, on the basis of the degree of similarity of their RNA sequences, living organisms could be grouped into three primary kingdoms. The first contained all of the typical bacteria, named by Woese eubacteria. A second group contained all species of higher plants and animals, while a third contained the relatively unknown methanogens.

Professor Woese speculated that methanogenic bacteria may have been the first bacteria on earth because of their capacity to thrive in anoxic conditions and their unique capacity to manufacture methane from H_2 and carbon dioxide; and he has therefore termed them archaebacteria. There is however, no basis for believing that the methanogens really are archaebacteria on the grounds of sequential comparisons. Because, just as in the case of classes derived from comparative protein sequences, none of the three classes, archaebacteria, eubacteria, or eucaryotes is intermediate with respect to the other classes, none of them can be designated ancestral or primitive with respect to the other.

The discovery that the methanogens belong to a quite new division of the living kingdom is yet another case which illustrates one of the main themes of this book, that whenever new types of organisms are occasionally discovered they never turn out to be ancestral to known groups but stand related only as sister groups in keeping with the thesis that nature's basic order is circumferential rather than sequential.

There is no evidence at all for evolutionary transformations in this sequencing data. The RNAs tell the same story as the proteins! On the basis of their RNA sequences, the three primary kingdoms stand equidistant apart, and equidistant from a theoretical common primeval ancestor.

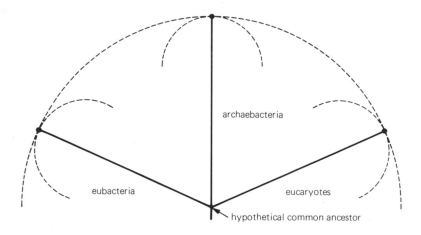

The above discussion has been concerned with comparisons of related molecules in different species. But there are many cases where a family of related proteins occur in the same species. For example, in man there are four related members of the haemoglobin family of proteins. There are three haemoglobins proper and of these α haemoglobin and β haemoglobin are found in the adult red blood cell while γ haemoglobin is found in the red cells of the fetus and newborn. Another member of the haemoglobin family is myoglobin which is found in both adult and fetal muscle cells and acts as an oxygen reservoir. These four proteins are very similar in terms of function, overall 3-D configuration, as well as amino acid sequence. The only evolutionary explanation that makes sense and has ever been proposed to account for their close similarities is that all four proteins originally evolved from a common precursor. The details and merits of the various evolutionary schemes that have been proposed need not concern us. The really significant finding that comes to light from comparing the proteins' amino acid sequences is that it is impossible to arrange them in any sort of evolutionary series.

Where there are four related protein sequences, say A B C D, a number of different evolutionary arrangements might be theoretically envisaged.

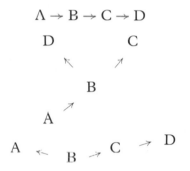

However, when the sequences of the haemoglobins in man are compared we find that myoglobin is quite distinct from the α β and γ haemoglobins, but unfortunately α β and γ haemoglobins are all equally distant from myoglobin. Further, when we compare the sequences of the α β and γ haemoglobins we find that, although β and γ are much closer together than either are to α, again the distances between β-γ and between α-γ are equal. We may classify the haemoglobins but the classification that results is the same groups-within-groups that we have seen when we compare different species at a molecular level.

Thousands of different sequences, protein and nucleic acid, have now been compared in hundreds of different species but never has any sequence been found to be in any sense the lineal descendant or ancestor of any other sequence. Anyone who doubts this need only consult the sequence difference matrices given in Dayhoff's standard reference book *Atlas of Protein Structure and Function*, available in any major library.

It is now well established that the pattern of diversity at a molecular level conforms to a highly ordered hierarchic system. Each class at a molecular level is unique, isolated and unlinked by intermediates. Thus molecules, like fossils have failed to provide the elusive intermediates so long sought by evolutionary biology. Again, the only relationships identified by this new technique are sisterly. At a molecular level, no organism is "ancestral" or "primitive" or "advanced" compared with its relatives. Nature seems to conform to the same non-evolutionary and intensely circumferential pattern that was long ago perceived by the great comparative anatomists of the nineteenth century.

One of the most remarkable features of these new biochemical discoveries is undoubtedly the way in which the pattern of molecular diversity seems to correspond to the predictions of typology. With very few exceptions the members of each defined taxa are always equally divergent whenever an outgroup comparison is made. Perhaps the only finding which does not seem to flow naturally from the typological model is that the degree of morphological divergence often does not seem to agree with the degree of molecular divergence. For example, the degree of molecular divergence among frogs, which are all morphologically very similar, is as great as that between mammals, which are morphologically very diverse.[8] Similarly, the proteins of conifers are as equally divergent as those of the flowering plants, a group which appears to be far more divergent than the conifers at a morphological level.[9] But despite those anomalies, all in all, the basic axioms of typology, that all the members of each type conform to type, that intratype variation is limited and type specific, so that when outgroup comparisons are made the subgroups of the type stand equidistant from more distantly related groups, hold universally throughout the entire realm of nature. This does not mean, of course, that typology is necessarily correct. But if we accept that closeness to empirical reality is the only criterion by which to judge alternative theories, we would, if strictly impartial, be forced to choose Aristotle and the eidos, in favour of Darwin and the theory of natural selection. There is little doubt that if this molecular evidence had been available one century ago it would have been seized upon with devastating effect by the opponents of evolution theory like Agassiz and Owen, and the idea of organic evolution might never have been accepted.

This new era of comparative biology illustrates just how erroneous is the assumption that advances in biological knowledge are continually confirming the traditional evolutionary story. There is no avoiding the serious nature of the challenge to the whole evolutionary framework implicit in these findings. For if the ancient representatives of groups such as amphibia, lungfish, cyclostomes and reptiles manufactured proteins similar to those manufactured by their living relatives today, and if, therefore, the isolation of the main divisions of nature was just the same in the past as it is today, if for example ancient lungfish and ancient amphibia were as separate from each other as their present day descendants are, then the whole concept of evolution collapses.

There are of course simply no objective grounds for excluding this possibility and, ironically, it was widely believed by most evolutionists before the full impact of these new comparative studies was realized. Thus, writing in the *Scientific American* in 1963, Zuckerkandl speculates:[10]

> Contemporary organisms that look much like ancient ancestral organisms probably contain a majority of polypeptide chains that resemble quite closely those of the ancient organisms. In other words, certain animals said to be "living fossils", such as the cockroach, the horseshoe crab, the shark and, among mammals, the lemur, probably manufacture a great many polypeptide molecules that differ only slightly from those manufactured by their ancestors millions of years ago.

The only way to save evolution in the face of these discoveries is to make the *ad hoc* assumption that the degree of biochemical isolation of the major groups was far less in the past, that ancient lungfish, for example, were far closer biochemically to ancient amphibia than their present day descendants. There is, however, absolutely no objective evidence that this assumption is correct. The only justification for such an assumption would be if evolution is true, but this is precisely the question at issue!

Given the distinctness of most of the divisions of nature at a morphological level and the absence of *bona fide* ancestors, intermediates or transitional forms, the credibility of evolutionary claims has had to depend traditionally very largely on "evidence" of a far from conclusive nature – on those instances where with the eye of faith it might be construed that, in Darwin's words, "a species or group like lungfish in some small degree connects by its affinities two large branches of life."[11] Thus the literature of biology is full of claims that this or that group, while not definitively intermediate or ancestral in any aspect of their biology, is at least so "in some degree" and its relationships with other groups may be interpreted in evolutionary terms.

So, according to Professor Romer, one of the leading vertebrate paleontologists:[12]

> The living cyclostomes and the fossil ostracoderms are members of a common stock of *primitive ancestral* vertebrates.
>
> [*emphasis added*]

and the opossums:[13]

> The opossums and their relatives found today in both Western continents are in almost every respect *ideal ancestors* for the whole marsupial group . . . In the late cretaceous of North America are found forms very similar to the living opossums.
>
> [*emphasis added*]

and the amphibians:[14]

> . . . are without question the *basal stock* from which the remaining group of land vertebrates have been derived.
>
> [*emphasis added*]

and among the placental mammals – the insectivores:[15]

> . . . There are still in existence a number of mammals, such as the shrews, moles and hedgehogs, which have retained . . . many *primitive characters*. These forms, grouped with related fossil types as the order Insectivora, are regarded as the most direct modern descendants of the primitive placentals.
>
> [*emphasis added*]

Similarly, despite the absence of clear-cut sequential arrangements, biologists have been able to allude to cases where nature does appear to fall approximately into a sequential pattern. One of the most celebrated cases of sequence is that of the vertebrate classes leading from the cyclostomes, through fish, amphibia and reptiles to the mammals. While no evolutionist has ever claimed that any of the living representatives of any vertebrate class is directly ancestral with respect to another vertebrate group, it is definitely implied that in terms of their general biology and overall morphology there are clear grounds for viewing the series as a natural phylogenetic sequence.

Potentially, comparative biochemistry by the demonstration of underlying sequential patterns could have added substantially to the credibility of such claims. If the sequence of vertebrate proteins could have been arranged in a series like the letter strings below:

```
Z ──────→ A ──────→ D        B           B
G ──────→ B        B ──────→ D           D
D        D ──────→ C        C           C
T        T        T ──────→ V ──────→ L
G ──────→ K        K ──────→ H           H
G        G        G        G ──────→ K
W        W ──────→ T        T ──────→ L
```

CYCLOSTOME → FISH → AMPHIBIAN → REPTILE → MAMMAL

then this would have provided powerful confirmation of the traditional sequential interpretation of the vertebrate classes. But as we have seen, the molecules provide little support for this "sequential" interpretation of the vertebrate classes.

In terms of their biochemistry, none of the species deemed "intermediate", "ancestral" or "primitive" by generations of evolutionary biologists, and alluded to as evidence of sequence in nature, shows any sign of their supposed intermediate status. The *cyclostomes are not primitive biochemically*; they are no closer to any non-vertebrate species than any other vertebrate group. The *opposums are not ideal ancestors* for the whole marsupial group in terms of their biochemistry. At a molecular level they are as far away from any reptile as any other marsupial species. Similarly, *the insectivores are no less indubitably mammalian* at a molecular level than any other mammalian group.

There is ultimately nothing contradictory in the molecular and morphological evidence. As we have already seen, to what extent the vertebrate sequence is actually supported by comparative anatomy is open to debate. There are morphological grounds for arranging the **vertebrate classes in a circumferential and typological arrangement which is perfectly in keeping with the molecular evidence.**

The fact that lungfish, monotremes and all the other favourite links of evolutionary biology give no hint of their supposed transitional status at a molecular level is perfectly in keeping with the fact that there are many morphological features of their biology which have never been easy to reconcile with their supposedly transitional status and which have always suggested that they represent unique and isolated types.

But by far the most challenging aspect of this new biochemical picture as far as evolution is concerned is the incredible orderliness of all the divisions. We have seen that the sequences of a protein such as haemoglobin vary considerably between the different members of a particular group but that when the sequences of any one group are compared with those of a more distantly related class the sequential divergence is invariably the same. The only way to explain this in evolutionary terms is to propose that since all the different lines of a group diverged each particular protein, such as haemoglobin or cytochrome C, has continued to evolve in each of the lines at its own characteristic uniform rate.

This can be seen by examining the hypothetical evolutionary tree below, where the nodes α β and γ represent presumed points of evolutionary divergence at times t_1, t_2 and t_3, and where the numbers indicate the sequential divergence between the various contemporary cytochromes sequences.

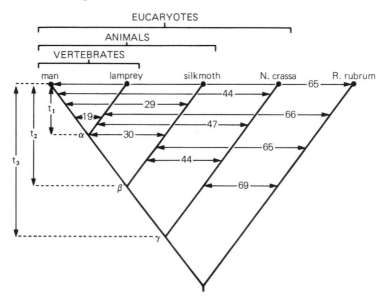

Consider first the evolution of the vertebrate cytochromes. Because they exhibit a considerable degree of sequential divergence (nineteen per cent), yet are equally isolated from non-vertebrate cytochromes, it must be presumed that both sequences have undergone the *same net degree of change in the same time interval t_1*. This means that since their common evolutionary divergence at node α, both cytochrome sequences must have diverged at a net constant rate with respect to absolute astronomical time.

Consider next the evolution of the animal cytochromes. Again because of the considerable degree of intra-group variation (as evidenced by the sequential divergence of twenty nine per cent between silk moth and man, and thirty per cent between silkmoth and lamprey), and because all three sequences are uniformly isolated from the two non-animal cytochromes, then it must be presumed that the three sequences have undergone the *same net degree of change in the same time interval t_2*. Once again, since their common evolutionary divergence at node β, the animal cytochromes must have diverged at a *net* constant rate with respect to absolute astronomical time.

Finally, consider the evolution of the eucaryotic cytochromes. Because of the amount of intra-group variation (as evidenced by the average forty five per cent sequential divergence between *H. crassa* and the three animal cytochromes), and because of the fact that all four sequences are almost exactly the same distance from their bacterial homologue, it must be presumed that all the eucaryotic cytochromes have undergone the same net degree of change in the same time interval t_3. Again since their common evolutionary divergence at node γ, all four cytochrome sequences must have diverged at a net constant rate with respect to absolute time.

We see then that the highly ordered pattern of cytochrome diversity could only have been generated if the overall net rate of sequential divergence had been constant with respect to absolute time in *all* the diverse branches of every class since their common evolutionary origin. Moreover, only if such a strange rule had been repeated over and over again, throughout eucaryotic evolution, following each evolutionary divergence, could it have generated the highly ordered pattern and uniform isolation of each class of eucaryotic cytochrome sequences.

Only if the degree of evolution in a family of molecules such as the cytochromes had been constrained by some kind of time constant mechanism, so that in any one class the degree of change which occurs is always proportional to the lapse of absolute time, can the ordered pattern of molecular diversity be explained. This remarkable concept is widely known as that of the 'molecular clock hypothesis'. But although such a clock is perfectly capable of accounting for the observed equal divergence of, say, all vertebrate cytochromes from those of insects, no one has been able to explain in precise terms exactly how such a time constant process could work. Rather than being a true explanation, the hypothesis of the molecular clock is really a tautology, no more than a restatement of the fact that at a molecular level the representatives of any one class are equally isolated from the representatives of another class.

The tautological nature of the molecular clock hypothesis is reminiscent of the explanations of the gaps in the fossil record. The proposal put forward to save evolution in the face of the missing links – that connecting links are missing from the fossil record because transitional species are very rare – is essentially tautological. If evolution is true then indeed the intermediates must be very rare. But unfortunately we can only know that evolution is true *after* we have found the transitional types! The explanation relies on belief in evolution in the first place. Similarly, if evolution is true then, yes indeed, the clock hypothesis must also be true. Again the hypothesis gets us nowhere. We save evolution because we believed it in the first place.

But there is an additional twist to the clock hypothesis. As we saw above, different proteins exhibit different degrees of interspecies variation. While haemoglobin sequences differ by fifty per cent between man and carp; cytochrome C differs by only thirteen per cent. To account for the fact that all the haemoglobin sequences of a particular group differ by fifty per cent from another group, while all the cytochrome C sequences differ by only thirteen per cent, it is necessary in evolutionary terms to presume that the molecular clock has ticked at a faster rate in the case of haemoglobin than in the case of cytochrome C; in other words, to propose two molecular clocks ticking at a different rate, one for the haemoglobin family and one for the cytochrome family. However, as there are hundreds of different families of proteins and each family exhibits its own unique degree of interspecies variation, some greater than haemoglobin, some far less than the cytochromes, then it is necessary to propose not just two clocks but one for each of the several hundred protein families, each ticking at its own unique and highly specific rate.

What sort of mutational mechanism might have generated uniform rates of evolution over vast periods of time in vastly dissimilar types of organisms? Basically, there are only two types of changes that can occur to the sequence of the genes specifying for functional proteins: neutral mutations which have no effect on function and are substituted by drift; and advantageous mutations which have a positive effect on function and are substituted by selection.

Unfortunately, neither evolution by genetic drift nor evolution by positive selection is likely to have generated anything remotely resembling a uniform rate of evolution in a family of homologous proteins.

The rate of genetic drift in any gene is directly related to and determined by the mutation rate. This is not controversial. The greater the mutation rate, the greater the speed of genetic drift. One fact that has perhaps lent a certain amount of support to the idea that drift might occur at approximately the same rate is the finding that in higher organisms the observed mutation rate per generation is approximately the same for many genes. The figure usually given for higher organisms is about 10^{-6}/gene/generation.[16] In closely related species such as man and chimpanzee, where the generation rates are similar, one might therefore expect approximately similar rates of drift to have occurred in homologous loci over several generations. But one would not expect similar rates of drift in more diverse types.

The proteins of small rodents, mice for example, are no more divergent than those in primates, elephants or whales, species which have very much longer generation times than rodents. A mouse may go through four to five generations in one year. The time taken by an elephant, a chimpanzee or a man to reach maturity is about fourteen, seven, ten years respectively. This means that at present the generation times of some mammalian species varies by a factor of nearly one hundred. Since the rodent order diverged from the primate, it is practically certain that the line leading to mouse has undergone nearly one hundred times as many reproductive cycles as that leading to man. If mutation rates are approximately constant per generation how then could drift have generated equal rates of genetic divergence in mice and men?

Among the insects there is an even greater diversity of generation times. A fruit fly may undergo a reproductive cycle in two weeks,[17] perhaps twenty generations per year. In the case of the Cicada one species has a generation cycle of seventeen years.[18] The generation rate of the Cicada is nearly one thousand times slower than that of fruit fly. The time of origin of the modern insect orders, families, and genera is not known, but many insect species are practically identical to the fossils found in Scandinavian amber some fifty million years ago.[19] If the differential generation times observed in modern species had only been maintained for as much as fifty million years, the fruit fly would have undergone fifty thousand million more generative cycles than the Cicada. Yet the proteins of different insect orders are equally divergent from those of vertebrates!

The plausibility of uniform drift shrinks even more when more diverse types of organisms are compared. Some higher plants, trees for example, only reach sexual maturity after eighty years.[20] Microorganisms such as yeast have generation times measured in minutes. The difference in generation times between such species is of the order of 10^5.

A uniform rate of drift in different lines is likely only if mutation rates in different organisms are uniform per unit time. This may be so in lines which have similar generation rates, such as man and chimpanzee or dog and fox. But many organisms often have vastly different generation times and all the evidence suggests that mutation rates per unit time are often very different in different species, varying by at least one to three orders of magnitude (see Figure 12.3).

Only if the rate of mutation in homologous proteins in different organisms, was for some mysterious reason adjusted so that it was constant with respect to absolute time would uniform rates of drift occur. As Ewens remarked at a recent symposium:[21]

> I note the well-known fact that the neutral theory predicts a constant rate of substitution per generation, whereas we appear to observe more a constant rate per year. In some of the species for which protein sequence comparisons have been made, there is a difference of one or even two orders of magnitude in generation time. It surely gets us nowhere simply to assume that the mutation rate adjusts itself in species of different generation time so that constant rates per year will arise.

Unfortunately, all the evidence suggests that in different groups of organisms the mutation rate per unit of absolute time is vastly different and this effectively excludes drift as a mechanism for the generation of uniform rates of evolution. On top of this there is the additional difficulty of envisaging how drift could have occurred at

ORGANISM	MUTATION RATE PER NUCLEOTIDE PER YEAR	
E. coli	0.7×10^{-6}	(a)
Drosophila	2.5×10^{-8}	(b)
Mouse	3.0×10^{-9}	(c)
Man	1.0×10^{-10}	(d)

Figure 12.3: Mutation Rates per Unit Time.

(a) assuming a mutation rate per base pair replication of 2×10^{-10} and 10 cell divisions per day.[22]

(b) assuming a mutation rate per cistron per generation of 10^{-6} that each cistron consists of 1000 nucleotides and a generation time of 2 weeks.[23]

(c) assuming a mutation rate per cistron per generation of 10^{-6} that each cistron consists of 1000 nucleotides and a generation time of 4 months.[24]

(d) assuming a mutation rate per cistron per generation of 10^{-6} that each cistron consists of 1000 nucleotides and a generation time of 10 years.[25]

different rates in different genes to account for the different rates of evolution in different families of homologous proteins.

One idea that has been put forward is that different proteins are under different functional constraints, which may have permitted some genes to have evolved faster than others. Alan Wilson, an authority in this field, wrote recently:[26]

> The proteins that evolve most slowly are supposed to have the highest proportion of sites at which the functional constraints are particularly severe. According to this view, nearly every mutation that could occur in the gene for histone 4* would be deleterious to the function of that histone.
>
> Conversely, the most rapidly evolving proteins are supposed to have the largest proportion of sites at which more than one residue would be

---

*The histones are a group of proteins which are intimately associated with the DNA in all eucaryotic, that is nucleated organisms which exhibit an astonishing invariance in their amino acid sequence. Histone 4 is the most invariant of all the histones.

compatible with function. Fibrinopeptides* are often cited as examples
of the latter type.

Like many explanations of phenomena which are on the face of
it difficult to reconcile with traditional evolutionary models, the
"functional constraints" hypothesis is largely tautological. Although
it is put forward as a solution to the problem of different rates of
protein evolution in different families of molecules, the only evidence
for the hypothesis is the observation it claims to explain.

Just because some vertebrate haemoglobins such as carp and man
differ from one another at up to eighty amino acid sites – while their
histones are identical, it cannot be inferred from this that the histones
are under more stringent selective constraints. Similarly, we do not
conclude that the selective pressures on vertebrate limbs are any less
intensive than those on vertebrate spinal columns merely because the
former exhibit much greater interspecies diversity than the latter.

Moreover, there is not a scrap of empirical evidence to suggest that
there is any systematic difference in the tolerance of different functional proteins to mutational change. As Wilson concedes:[27]

> ... we are not aware of direct experimental evidence showing rigorously
> that histone function is especially sensitive to amino acid substitution
> or that fibrinopeptide function is especially insensitive to amino acid
> substitution. Experimental studies would require that quantitative in
> vitro assays for the specific functions of histone 4 and fibrinopeptides
> be available. These have not been developed for histones, fibrinopep-
> tides, or, indeed, most of the proteins whose evolutionary rates are
> listed.

Further theoretical arguments can be advanced against the idea.
Most of the functional criteria which must be satisfied in amino acid
sequences, for example, those related to protein stability and the
necessity for folding algorithms (see Chapter Ten), are of a general
nature and are unlikely to differ in different proteins or in different
species or at different times in the past.

The degree of stringency of the criteria for protein stability, for
example, is almost certainly the same in all existing proteins and was
probably the same in blue green algae 3,500 million years ago unless,

---

*The fibrinopeptides are two short amino acid sequences which are removed from
the protein fibrinogen during the process of blood coagulation.

of course, one is prepared to presume that the basic physical constants have changed during geological time so that the nature of weak chemical interactions, hydrogen bonds etc, and their influence on the stability of α helices and β pleated sheet* formations are different today than in the past.

Similarly, the criteria for function of the active sites of enzymes could hardly vary much from protein to protein. *All* active sites depend on an exact atomic fit between the substrate and protein molecule. It is difficult to see why the tolerance levels should vary significantly.

Considering the vast complexity of all gene sequences it seems extremely unlikely that the functional constraints operating on the great majority of proteins, apart possibly from a few "junk" spacer sequences, vary according to any sort of systematic pattern. It is much, very much, more likely that the overall constraints on most protein sequences are somewhat similar in different proteins and in different species and would have been so over hundreds of millions of years of evolution.

Again, it is the sheer universality of the phenomenon – the necessity to believe that the functional constraints in *all* the members of a particular protein family, say A, in *all* diverse phylogenetic lines for *all* of hundreds of millions of years have remained precisely five times as stringent as those operating on the members of another protein family, say B – which fatally weakens the theory.

But if neutral drift gets us nowhere, selectionist explanations fare no better. It is very difficult to understand why all the members of a particular family of proteins, such as the haemoglobins or the cytochromes, should have suffered the same number of advantageous mutations since their common divergence. Selectionist explanations are particularly implausible in the case of the living fossils. While most species make only, what is on a geological time scale, a fleeting appearance in the fossil record, often no more than a few million years, some have persisted almost unchanged for hundreds of millions of years down to modern times – these we call the living fossils.

These great survivors have always held a great fascination for scientists because they are biological time capsules, preserving in their morphology, physiology and behaviour a pattern of life from

---

* α helices and β pleated sheet conformations are highly ordered configurations adopted by the folded amino acid sequences in many different kinds of proteins.

the remote past, but in the context of the molecular clock hypothesis they have taken on an added significance. The lungfish is a classic example. These remarkable fish are found in the swamps, rivers and lakes of central Africa. As well as gills the lungfish also possesses an efficient lung which it uses to survive during the severe droughts which periodically afflict the African plains. At the onset of the dry season, as the depth of the water decreases and while the mud is still soft enough for burrowing, the fish digs itself a bulbous cavity that opens to the surface through a small hole and there, just a few feet beneath the parched surface, curled up and encased in its cocoon of mud, it lies dormant. Over the months it slowly becomes severely dehydrated; its skin dries and wrinkles so that it has a lifeless mummified appearance; but as soon as the waters return it emerges from its tomb, takes on water, and returns to its active existence as a fish.

The modern lungfishes are members of an ancient group of fishes which are closely related to the rhipidistian fishes, the group considered almost directly ancestral to the amphibia. Lungfish almost identical to those of modern Africa are found as fossils in the rocks of the Devonian era 350,000,000 years ago alongside fossils of the earliest amphibians and the very fish groups from which the amphibia supposedly arose. Through millions of years since Devonian times, uninfluenced by all the massive changes in the Earth's crust and fauna, while the ancient super-continents of Gonwanaland and Laurentia fragmented, while the dinosaurs came and went, the lungfishes continued performing their unique ritual of survival.

In evolutionary terms the lungfish and other living fossils are in a very real sense like samples drawn an eternity ago from near the main course of the stream of vertebrate life. While the tree of evolution continued to grow in all directions above them, they remained the same, so that over the eons of time they were increasingly left behind, increasingly primitive and ancestral with respect to the newer groups.

Yet the proteins of lungfish are just as far from lamprey as any other fish, amphibian or mammalian group! If we are to explain this in terms of selection we must presume that the proteins of this living fossil have been subject to the same net rate of advantageous amino acid replacement over four hundred million years as the proteins of organisms which have been morphologically transformed out of all recognition over the same period of time. But this is verging on *reductio ad absurdum* because it necessitates a complete divorce between adaptive change at a molecular and at a morphological level.

Consider the case of the haemoglobin in man and lungfish. Since the two lines are presumed to have diverged in Devonian times, some four hundred million years ago, the line leading to man has undergone profound physiological and morphological changes, while the modern lungfish is still very close in terms of its morphology and physiology to the ancient fishes. The line leading to man has supposedly undergone three fundamental transformations, the amphibian, the amniotic, and the mammalian. During the course of these presumed transformations the cardiovascular system has undergone enormous and dramatic changes. The heart has changed from a simple tube-like organ to a four-chambered efficient pump. The gills and branchial arteries have been replaced by lungs and the pulmonary circulation. The system of oxygenation has been utterly transformed. At the same time, the red blood cells themselves have become completely different. From the large round red cells of diameter of approximately $20\,\mu$ ($\mu = 10^{-6}$ meter) typical biochemically of relatively unspecialized eucaryotic cells possessing nucleus, mitochondria etc, they have changed into small plate-like structures of diameter $7\,\mu$ without nucleus or mitochondria and containing very much more haemoglobin per unit volume.

While this dramatic series of morphological, physiological, histological and biochemical changes were supposedly occurring along the lineage leading to the mammals, the morphological, physiological, and histological organization of the cardiovascular system of the line leading to lungfish must have remained virtually unchanged.

It is very difficult to understand why a protein functioning in the basically unchanging physiological environment of the lungfish's red cell should have undergone precisely the same number of beneficial mutations as a related protein evolving in a line subject to such global adaptational changes. While selection at the morphological and molecular level may be relatively unlinked, it is surely inconceivable that they could be *absolutely* unrelated. All the biology of an organism, all its anatomical features, its physiological and metabolic functions are ultimately reducible to its constituent proteins. Because organisms are systematic wholes in which every component more or less interacts with every other component, because all the functional components of living things are all ultimately made up of proteins, then inevitably every physiological or structural change is bound to impinge on the functionality of proteins. Proteins cannot be isolated from the environment in which they function.

Unfortunately, the case of lungfish haemoglobin is not unique. The opossum is another classic living fossil, virtually unchanged morphologically from its ancient ancestors of the late Cretaceous period nearly one hundred million years ago. But when opossum haemoglobin is compared with the haemoglobins of other mammals it is in no way primitive with respect to other mammalian species. In fact, rather the reverse; if anything, opossum haemoglobin is actually slightly further away from presumed common ancestors of mammals such as fish and amphibia than other mammalian species. So this mammalian species, a living fossil, apparently unchanged morphologically for nearly a hundred million years, a species which predates the entire adaptive radiation of the placental orders, has a haemoglobin as far removed from presumed mammalian ancestors as any of the recently evolved mammalian types!

Of course, the implausibility of selectionist explanations do not stop with the haemoglobins of a few living fossils. As in the case of uniform drift it is the sheer universality of the phenomenon – the necessity to believe that since their common divergence every single family of homologous proteins have suffered the same number of adaptive substitutions over the same period of time in *all* phylogenetic lines – which fatally weakens selectionist explanations.

Perhaps one of the most difficult problems in this whole area is trying to provide an explanation of how a uniform rate of evolution could have occurred in amino acid sequences which apparently perform no function other than acting as spacer sequences linking together the functional regions of a protein. A classic example of this are two short amino acid sequences which are snipped out of the protein fibrinogen after it is activated during blood coagulation. These are known as fibrinopeptides A and B. As far as is known, neither of these two short peptides have any biological function, yet their percent sequence divergence in different mammalian groups conforms to the same ordered pattern as is found in all other proteins, ie, the fibrinopeptides in all the members of any group are equally isolated from all the fibrinopeptide sequences found outside their group. If we are to explain this in terms of evolution we must again assume that an equal degree of fibrinopeptide sequential change has occurred in all the diverse lines of a particular group since their common divergence.

If such sequences really are under no selective constraints then drift is the only agent that could have been responsible for the pattern of interspecies differences. Neutral sequences are by definition outside the surveillance of natural selection but this leads to a serious dilemma. As we have seen above, there is no conceivable way in which a uniform rate of drift could have occurred in organisms as diverse as mouse and man and yet the fibrinopeptides in rodents are isolated to exactly the same degree as those in primates. Drift seems to be excluded.

But selectionist explanations seem to lead to absurd conclusions. Because the spacer sequences such as the fibrinopeptides exhibit the highest interspecies divergence of all proteins, if this is to be accounted for on purely selectionist grounds it is necessary to propose that they must have suffered adaptive changes very much more often than proteins such as the haemoglobins or the cytochromes. In other words, they must have been under the intense scrutiny of natural selection. Not only must such sequences have suffered more adaptive changes than other proteins but in addition, these substitutions must have occurred regularly.

The difficulties associated with attempting to explain how a family of homologous proteins could have evolved at constant rates has created chaos in evolutionary thought. The evolutionary community has divided into two camps – those still adhering to the selectionist position, and those rejecting it in favour of the neutralist. The devastating aspect of this controversy is that neither side can adequately account for the constancy of the rate of molecular evolution, yet each side fatally weakens the other. The selectionists wound the neutralists' position by pointing to the disparity in the rates of mutation per unit time, while the neutralists destroy the selectionist position by showing how ludicrous it is to believe that selection would have caused equal rates of divergence in "junk" proteins or along phylogenetic lines so dissimilar as those of man and carp. Both sides win valid points, but in the process the credibility of the molecular clock hypothesis is severely strained and with it the whole paradigm of evolution itself is endangered.

There is simply no way of explaining how a uniform rate of evolution could have occurred in any family of homologous proteins by either chance or selection; and, even if we could advance an explanation for one particular protein family, we would still be left with the mystifying problem of explaining why other protein families should have evolved at different rates. The more deeply the problem is examined the less it appears amenable to solution in terms of chance and selection.

Despite the fact that no convincing explanation of how random evolutionary processes could have resulted in such an ordered pattern of diversity, the idea of uniform rates of evolution is presented in the literature as if it were an empirical discovery. The hold of the evolutionary paradigm is so powerful that an idea which is more like a principle of medieval astrology than a serious twentieth-century scientific theory has become a reality for evolutionary biologists.

Here is, perhaps, the most dramatic example of the principle that wherever we find significant empirical discontinuities in nature we invariably face great, if not insurmountable, conceptual problems in envisaging how the gaps could have been bridged in terms of gradual random processes. We saw this in the fossil record, we saw it in the case of the feather, in the case of the avian lung and in the case of the wing of the bat. We saw it again in the case of the origin of life and we see it here in this new area of comparative biochemistry.

What has been revealed as a result of the sequential comparisons of homologous proteins is an order as emphatic as that of the periodic table. Yet in the face of this extraordinary discovery the biological community seems content to offer explanations which are no more than apologetic tautologies.

NOTES

1. Dayhoff, M. D. (1972) *Atlas of Protein Sequence and Structure*, National Bio-medical Research Foundation, Silver Spring, Maryland, vol 5, Matrix 1, p D–8.
2. ibid, Matrix 1, p D–8.
3. ibid, Matrix 1, p D–8.
4. ibid, Matrix 7, p D–52.
5. ibid, Matrix 1, p D–8; and Matrix 12, p D–88.
6. ibid, Matrix 10, p D–56; and Matrix 12, p D–88.
7. Woese, C. R. and Fox, G. E. (1977) "Phylogenetic Structure of the Prokaryotic Domain: The Primary Kingdoms", *Proc. Natl. Acad. Sci. U.S.*, 74: 5088–90.
8. Ferguson, A. (1980) *Biochemical Systematics and Evolution*, Blackie, Glasgow, p 156–57.
9. Prager, E. M., Fowler, D. P. and Wilson, A. C. (1976) "Rates of Evolution in Conifers (Pinaceae)", *Evolution* 30: 637–49.
10. Zukerkandl, E. (1965) "The Evolution of Haemoglobin", *Scientific American*, 213(5): 110–18, see p111.
11. Darwin, C. (1872) *The Origin of Species*, 6th ed (1962) Collier Books, New York, p 137.

12. Romer, A.S. (1966) *Vertebrate Paleontology*, 3rd ed, University of Chicago Press, Chicago, p15.
13. ibid, p204.
14. ibid, p78.
15. ibid, p208.
16. Drake, J.W. (1970) *The Molecular Basis of Mutation*, Holden-Day Inc, San Francisco, Table 5–3. Sager, R. and Ryan, F.J. (1961) *Cell Heredity*, John Wiley and Sons, Inc, Table 2.3, p55. Cavalli-Sforza, L.L. and Bodmer, W.F. (1971) *The Genetics of Human Populations*, Freedman and Co, San Francisco, p108.
17. Demerec, M. and Kaufmann, B.D. (1961) *Drosophila Guide*, 7th ed, Carnegie Institute of Washington, p3.
18. Imms, A.D. (1957) *A General Textbook of Entomology*, Methuen and Co Ltd, London, p439.
19. Brues, C.T. (1951) "Insects in Amber", Scientific American, 185(5), pp56–61.
20. Thomson, J.A. (1934) *Biology for Everyman*, 2vols, J.M. Dent and Sons Ltd, vol2, p1241.
21. Ewens, W.J. (1973) "Comments on Dr Kimura's Paper", Genetics Supplement, *Genetics*, 73: 36–38, p36.
22. Drake, op cit.
23. ibid.
24. Cavalli-Sforza, op cit.
25. ibid.
26. Wilson, A.C. Carlson, S.S. and White, T.J. (1977), "Biochemical Evolution", *Ann. Rev. Biochem.*, 46: 573–639, see p611.
27. ibid.

CHAPTER 13
Beyond the Reach of Chance

> He who believes that some ancient form was transformed suddenly . . .
> will further be compelled to believe that many structures beautifully
> adapted to all the other parts of the same creature and to the sur-
> rounding conditions, have been suddenly produced . . . To admit all
> this is, as it seems to me, to enter into the realms of miracle, and to
> leave those of Science.

According to the central axiom of Darwinian theory, the initial
elementary mutational changes upon which natural selection acts are
entirely random, completely blind to whatever effect they may have
on the function or structure of the organism in which they occur,
"drawn", in Monod's words,[1] from "the realm of pure chance". It is
only after an innovation has been disclosed by chance that it can be
seen by natural selection and conserved.

Thus if follows that every adaptive advance, big or small, discovered during the course of evolution along every phylogenetic line
must have been found as a result of what is in effect a purely random
search strategy. The essential problem with this "gigantic lottery"
conception of evolution is that all experience teaches that searching
for solutions by purely random search procedures is hopelessly
inefficient.

Consider first the difficulty of finding by chance English words
within the infinite space of all possible combinations of letters. A
section of this space would resemble the following block of letters:

```
FLNWCYTQONMCDFEUTYLDWPQXCZNMIPQZXHGOT
IRJSALXMZVTNCTDHEKBUZRLHAJCFPTQOZPNOTJXD
WHYGCBZUDKGTWIBMZGPGLAOTDJZKXUEMWBCNX
YTKGHSBQJVUCPDLWKSMYJVGXUZIEMTJBYGLMPSJS
KFURYEBWNQPCLXKZUFMTYBUDISTABWNCPDORIS
MXKALQJAUWNSPDYSHXMCKFLQHAVCPDYRTSIZSJR
YFMAHZLVPRITMGYGBFMDLEPE
```

Within the total letter space would occur every single English word and every single English sentence and indeed every single English book that has been or will ever be written. But most of the space would consist of an infinity of pure gibberish.

Simple three letter English words would be relatively common. There are $26^3 = 17,000$ combinations three letters long, and, as there are about five hundred three letter words in English, then about one in thirty combinations will be a three letter word. All other three letter combinations are nonsense. To find by chance three letter words, eg "not", "bud", "hut", would be a relatively simple task necessitating a search through a string of only about thirty or so letters.

Because three letter words are so probable it is very easy to go from one three letter word to another by making random changes to the letter string. In the case of the word "hat" for example, by randomly substituting letters in the position occupied by h in the word we soon hit on a new three letter word:

> hat
> aat
> bat
> cat
> dat
> eat
> fat

Thus not only is it possible to find three letter English words by chance but because the probability gaps between them are small, it is easy to transform any word we find into a quite different word through a sequence of probable intermediates:

$$hat \rightarrow cat \rightarrow can \rightarrow tan \rightarrow tin$$

However, to find by chance longer words, say seven letters long such as "English" or "require", would necessitate a vastly increased search. There are 26^7 or 10^9, that is, one thousand million combinations of letters seven letters long. As there are certainly less than ten thousand English words seven letters long, then to find one by chance we would have to search through letter strings in the order of one hundred thousand units long. Twelve letter words such as "construction" or "unreasonable" are so rare that they occur only once by chance in strings of letters 10^{14} units long; as there are about 10^{14} minutes in one thousand million years one can imagine how long a monkey at a typewriter would take to type out by chance one English word twelve letters long. Intuitively it seems unbelievable that such apparently simple entities as twelve letter English words could be so rare, so inaccessible to a random search.

The problem of finding words by chance arises essentially because the space of all possible letter combinations is immense and the overwhelming majority are complete nonsense; consequently meaningful sequences are very rare and the probability of hitting one by chance is exceedingly small.

Moreover, even if by some lucky fluke we were to find, say, one twelve letter word by chance, because each word is so utterly isolated in a vast ocean of nonsense strings it is very difficult to get another meaningful letter string by randomly substituting letters and testing each new string to see if it forms an English word. Take the word "unreasonable". There are a few closely related words such as "reasonable," "reason," "season," "treason," or "able," which can be reached by making changes to the letter sequence but the necessary letter changes are unfortunately highly specific and finding them by chance involves a far longer and more difficult search than was the case with three letter words.

Sentences, of course, even short ones, are even rarer and long sentences rare almost beyond imagination. Linguists have estimated a total of 10^{25} possible English sentences one hundred letters long, but as there are a total of 26^{100} or 10^{130} possible sequences one hundred letters long, then less than one in about 10^{100} will be an English sentence. The figure of 10^{100} is beyond comprehension – some idea of the immensity it represents can be grasped by recalling that there are only 10^{70} atoms in the entire observable universe.

Each English sentence is a complex system of letters which are integrated together in highly specific ways: firstly into words, then into word phrases, and finally into sentences. If the subsystems are all to be combined in such a way that they will form a grammatical English sentence than their integration must follow rigorously the *a priori* rules of English grammar. For example, one of the rules is that the letters in the sentence must be combined in such a way that they form words belonging to the lexicon of the English language.

However, random strings of English words, eg "horse", "cog", "blue", "fly", "extraordinary", do not form sentences because there exists a further set of rules – the rules of syntax which dictate, among other things, that a sentence must possess a subjective and a verbal clause.

On top of this there exists a further set of rules which governs the semantic relationship of the components of a sentence. Obviously, not all strings of English words which are arranged correctly according to the rules of English syntax are meaningful. For example: "The raid (subject) ate (verb) the sky (object)." Each word is from the English lexicon and their arrangement satisfies the rules of syntax. However, the sequence disobeys the rules of semantics and is as nonsensical as a completely random string of letters.

The rules of English grammar are so stringent that only highly specialized letter combinations can form grammatical sentences and consequently, because of the immensity of the space of all possible letter combinations, such highly specialized strings are utterly lost within it, infinitely rare and isolated, absolutely beyond the reach of any sort of random search that could be conceivably carried out in a finite time even with the most advanced computers on earth. Moreover, because sentences are so rare and isolated, even if one was discovered by chance the probability gap between it and the nearest related sentence is so immeasurable that no conceivable sort of random change to the letter or word sequence will ever carry us across the gap.

Consider the sentence: "Because of the complexity of the rules of English grammar most English sentences are completely isolated." If we set out to reach another sentence by randomly substituting a new word in place of an existing word and then testing the newly created sequence of words to see if it made a grammatical sentence, we would find very few substitutions were grammatically acceptable and even to find *one* grammatical substitution would take an unbelievably long time if we searched by pure chance. Some of the few grammatical substitutions are shown below:

because	English
of	grammar
the	most → all → some
complexity → nature	english
of	sentences
the	are
rules → algorithms	completely → totally → invariably → always
of	isolated → immutable

There are about 10^5 words in the lexicon of the English language and, as there are sixteen words in the above sentence, we would have $1·6 \times 10^6$ possibilities to test. If there are, say, two hundred individual words out of the $1·6 \times 100^6$ which can be substituted grammatically, we would have to test about eight thousand words on average before we found a grammatical substitution.

Testing one new word per minute, it would take us five days working day and night to find by chance our first grammatical substitution, and to test all the possible words in every position in the sentence would take about three years, and after three years of searching all we would have achieved would be a handful of sentences closely related to the one with which we started.

Sentences are not the only complex systems which are beyond the reach of chance. The same principles apply, for example, to watches, which are also highly improbable, and where consequently each different functional watch is intensely isolated by immense probability gaps from its nearest neighbours.

To see why, we must begin by trying to envisage a universe of mechanical objects containing all possible combinations of watch components: springs, gears, levers, cogwheels, each of every conceivable size and shape. Such a universe would contain every functional watch that has ever existed on earth and every functional watch that could possibly exist at any time in the future. Although we cannot in this case calculate the rarity of functional combinations (watches that work) as we could in the case of words and sentences, common sense tells us that they would be exceedingly rare. Our imaginary universe would mostly consist of combinations of gears and cogwheels which would be entirely useless; each functional watch would, like a meaningful sentence, be an isolated island separated from all other islands of function by a surrounding infinity of junk composed only of incoherent and functionless combinations.

Again, as with sentences, because the total number of incoherent nonsense combinations of components vastly exceeds, by an almost inconceivable amount, the tiny fraction which can form coherent combinations, function is exceedingly rare. If we were to look by chance for a functional watch we would have to search for an eternity amid an infinity of combinations until we hit upon a functional watch.

The basic reason why functional watches are so exceedingly improbable is because, to be functional, a combination of watch components must satisfy a number of very stringent criteria (equivalent to the rules of grammar), and these can only be satisfied by highly specialized unique combinations of components which are coadapted to function together. One rule might be that all cogwheels must possess perfect regularly-shaped cogs; another rule might be that all the cogs must fit together to allow rotation of one wheel to be transmitted throughout the system.

It is obviously impossible to contemplate using a random search to find combinations which will satisfy the stringent criteria which govern functionality in watches. Yet, just as a speaker of a language cognizant with the rules of grammar can generate a functional sentence with great ease, so too a watchmaker has little trouble in assembling a watch by following the rules which govern funtionality in combinations of watch components.

What is true of sentences and watches is also true of computer programs, airplane engines, and in fact of all known complex systems. Almost invariably, function is restricted to unique and fantastically improbable combinations of subsystems, tiny islands of meaning lost in an infinite sea of incoherence. Because the number of nonsense combinations of component subsystems vastly exceeds by unimaginable orders of magnitude the infinitesimal fraction of combinations in which the components are capable of undergoing coherent or meaningful interactions. Whether we are searching for a functional sentence or a functional watch or the best move in a game of chess, the goals of our search are in each case so far lost in an infinite space of possibilities that, unless we guide our search by the use of algorithms which direct us to very specific regions of the space, there is no realistic possibility of success.

Discussing a well known checker-playing program, Professor Marvin Minsky of the Massachusetts Institute of Technology comments:[2]

> This game exemplifies the fact that many problems can in principle be solved by trying all possibilities – in this case exploring all possible moves, all the opponent's possible replies, all the player's possible replies to the opponent's replies and so on. If this could be done, the player could see which move has the best chance of winning. In practice, however, this approach is out of the question, even for a computer; the tracking down of every possible line of play would involve some 10^{40} different board positions. A similar analysis for the game of chess would call for some 10^{120} positions. Most interesting

problems present far too many possibilities for complete trial and error analysis.

Nevertheless, as he continues, a computer can play checkers if it is capable of making intelligent limited searches:

> Instead of tracking down every possible line of play the program uses a partial analysis (a "static evaluation") of a relatively small number of carefully selected features of a board position – how many men there are on each side, how advanced they are and certain other simple relations. This incomplete analysis is not in itself adequate for choosing the best move for a player in a current position. By combining the partial analysis with a limited search for some of the consequences of the possible moves from the current position, however, the program selects its move as if on the basis of a much deeper analysis. The program contains a collection of rules for deciding when to continue the search and when to stop. When it stops, it assesses the merits of the "terminal position" in terms of the static evaluation. If the computer finds by this search that a given move leads to an advantage for the player in all the likely positions that may occur a few moves later, whatever the opponent does, it can select this move with confidence.

The inability of unguided trial and error to reach anything but the most trivial of ends in almost every field of interest obviously raises doubts as to its validity in the biological realm. Such doubts were recently raised by a number of mathematicians and engineers at an international symposium entitled "Mathematical Challenges to the Neo-Darwinian Interpretation of Evolution",[3] a meeting which also included many leading evolutionary biologists. The major argument presented was that Darwinian evolution by natural selection is merely a special case of the general procedure of problem solving by trial and error. Unfortunately, as the mathematicians present at the symposium such as Schutzenberger and Professor Eden from MIT pointed out, trial and error is totally inadequate as a problem solving technique without the guidance of specific algorithms, which has led to the consequent failure to simulate Darwinian evolution by computer analogues. For similar reasons, the biophysicist Pattee has voiced scepticism over natural selection at many leading symposia over the past two decades. At one meeting entitled "Natural Automata and Useful Simulations", he made the point:[4]

Even some of the simplest artificial adaptive problems and learning games appear practically insolvable even by multistage evolutionary strategies.

Living organisms are complex systems, analogous in many ways to non-living complex systems. Their design is stored and specified in a linear sequence of symbols, analogous to coded information in a computer programe. Like any other system, organisms consist of a number of subsystems which are all coadapted to interact together in a coherent manner: molecules are assembled into multimolecular systems, multimolecular assemblies are combined into cells, cells into organs and organ systems finally into the complete organism. It is hard to believe that the fraction of meaningless combinations of molecules, of cells, of organ systems, would not vastly exceed the tiny fraction that can be combined to form assemblages capable of exhibiting coherent interactions. Is it really possible that the criteria for function which must be satisfied in the design of living systems are at every level far less stringent than those which must be satisfied in the design of functional watches, sentences or computer programs? Is it possible to design an automaton to construct an object like the human brain, laying down billions of specific connections, without having to satisfy criteria every bit as exacting and restricting as those which must be met in other areas of engineering?

Given the close analogy between living systems and machines, particularly at a molecular level, there cannot be any objective basis to the assumption that functional organic systems are likely to be less isolated or any easier to find by chance. Surely it is far more likely that functional combinations in the space of all organic possibilities are just as isolated, just as rare and improbable, just as inaccessible to a random search and just as functionally immutable by any sort of random process. The only warrant for believing that functional living systems are probable, capable of undergoing functional transformation by random mechanisms, is belief in evolution by the natural selection of purely random changes in the structure of living things. But this is precisely the question at issue.

If complex computer programs cannot be changed by random mechanisms, then surely the same must apply to the genetic programmes of living organisms. The fact that systems in every way analogous to living organisms cannot undergo evolution by pure trial and error and that their functional distribution invariably conforms to an improbable discontinuum comes, in my opinion, very close to a formal disproof of the whole Darwinian paradigm of nature. By what strange capacity do living organisms defy the laws of chance which are apparently obeyed by all analogous complex systems?

We now have machines which exhibit many properties of living systems. Work on artificial intelligence is advanced and the possibility of constructing a self-reproducing machine was discussed by the mathematician von Neumann in his now famous *Theory of Self-Reproducing Automata.*[5] Although some advanced machines can solve simple problems, none of them can undergo evolution by the selection of random changes in their structure without the guidance of already existing programmes. The only sort of machine that might, at some future date, undergo some sort of evolution would be one exhibiting artificial intelligence. Such a machine would be capable of altering its own organization in an intelligent way. However evolution of this sort would be more akin to Lamarckian, but by no stretch of the imagination could it be considered Darwinian. The construction of a self-evolving intelligent machine would only serve to underline the insufficiency of unguided trial and error as a causal mechanism of evolution.

It was the close analogy between living systems and complex machines and the impossiblity of envisaging how objects could have been assembled by chance that led the natural theologians of the eighteenth and early nineteenth centuries to reject as inconceivable the possibility that chance would have played any role in the origin of the complex adaptations of living things. William Paley, in his classic analogy between an organism and a watch, makes precisely this point:[6]

> Nor would any man in his senses think the existence of the watch, with its various machinery, accounted for, by being told that it was one out of possible combinations of material forms; that whatever he had found in the place where he found the watch, must have contained some internal configuration or other; and that this configuration might be the structure now exhibited, viz. of the works of a watch, as well as a different structure.

It is true that some authorities have seen an analogy to evolution by natural selection in gradual technological advances. Jukes, for instance, in a recent letter to *Nature* drew an analogy between the evolution of the Boeing 747 from Bleriots' 1909 monoplane through the Boeing Clippers in the 1930s to the first Boeing jet airliner, the 707, which started in service in 1959 and which was the immediate predecessor of the 747s, and biological evolution. In his words:[7]

> The brief history of aircraft technology is filled with branching processes, phylogeny and extinctions that are a striking counterpart of three billion years of biological evolution.

Unfortunately, the analogy is false. At no stage during the history of the aviation industry was the design of any flying machine achieved by chance, but only by the most rigorous applications of all the rules which govern function in the field of aerodynamics. It is true, as Jukes states, that "wide-bodied jets evolved from small contraptions made in bicycle shops, or in junkyards," but they did not evolve by chance.

There is no way that a purely random search could ever have discovered the design of an aerodynamically feasible flying machine from a random assortment of mechanical components – again, the space of all possibilities is inconceivably large. All such analogies are false because in *all* such cases the search for function is intelligently guided. It cannot be stressed enough that evolution by natural selection is analogous to problem solving without any intelligent guidance, without any intelligent input whatsoever. No activity which involves an intelligent input can possibly be analogous to evolution by natural selection.

The above discussion highlights one of the fundamental flaws in many of the arguments put forward by defenders of the role of chance in evolution. Most of the classic arguments put forward by leading Darwinists, such as the geneticist H. J. Muller and many other authorities including G. G. Simpson, in defence of natural selection make the implicit assumption that islands of function are common, easily found by chance in the first place, and that it is easy to go from island to island through functional intermediates.

This is how Simpson, for example, envisages evolution by natural selection:[8]

> How natural selection works as a creative process can perhaps best be explained by a very much oversimplified analogy. Suppose that from a pool of all the letters of the alphabet in large, equal abundance you tried to draw simultaneously the letters *c*, *a*, and *t*, in order to achieve a purposeful combination of these into the word "cat". Drawing out

three letters at a time and then discarding them if they did not form this useful combination, you obviously would have very little chance of achieving your purpose. You might spend days, weeks, or even years at your task before you finally succeeded. The possible number of combinations of three letters is very large and only one of these is suitable for your purpose. Indeed, you might well never succeed, because you might have drawn all the *c*'s, *a*'s, or *t*'s in wrong combinations and have discarded them before you succeeded in drawing all three together. But now suppose that every time you draw a *c*, an *a*, or a *t* in a wrong combination, you are allowed to put these desirable letters back in the pool and to discard the undesirable letters. Now you are sure of obtaining your result, and your chances of obtaining it quickly are much improved. In time there will be only *c*'s, *a*'s, and *t*'s in the pool, but you probably will have succeeded long before that. Now suppose that in addition to returning *c*'s, *a*'s, and *t*'s to the pool and discarding all other letters, you are allowed to clip together any two of the desirable letters when you happen to draw them at the same time. You will shortly have in the pool a large number of clipped *ca*, *ct*, and *at* combinations plus an also large number of the *t*'s, *a*'s, and *c*'s needed to complete one of these if it is drawn again. Your chances of quickly obtaining the desired result are improved still more, and by these processes you have "generated a high degree of improbability" – you have made it probable that you will quickly achieve the combination *cat*, which was so improbable at the outset. Moreover, you have created something. You did not create the letters *c*, *a*, and *t*, but you have created the word "cat," which did not exist when you started.

The obvious difficulty with the whole scheme is that Simpson assumes that finding islands of function in the first place (the individual letters *c*, *a*, and *t*) is highly probable and that the functional island "cat" is connected to the individual letter islands by intermediate functional islands *ca*, *ac*, *ct*, *at*, *ta*, *fc*, so that we can cross from letters to islands by natural selection in unit mutational steps In other words, Simpson has assumed that islands of function are very probable, but this is the very assumption which must be proved to show that natural selection would work.

Obviously, if islands of function in the space of all organic possibilities are common, like three or four letter words, then of course functional biological systems will be within the reach of chance; and because the probability gaps will be small, random mutations will easily find a way across. However, as is evident from the above discussion, Simpson's scheme, and indeed the whole Darwinian framework, collapse completely if islands of function are like twelve letter words or English sentences.

These considerations of the likely rarity and isolation of functional systems within their respective total combinatorial spaces also reveals the fallacy of the current fashion of turning to saltational models of evolution to escape the impasse of gradualism. For as we have seen, in the case of every kind of complex functional system the total space of all combinatorial possibilities is so nearly infinite and the isolation of meaningful systems so intense, that it would truly be a miracle to find one by chance. Darwin's rejection of chance saltations as a route to new adaptive innovations is surely right. For the combinatorial space of all organic possibilities is bound to be so great, that the probability of a sudden macromutational event transforming some existing structure or converting *de novo* some redundant feature into a novel adaptation exhibiting, that "perfection of coadaptation" in all its component parts so obvious in systems like the feather, the eye or the genetic code and which is necessarily ubiquitous in the design of all complex functional systems biological or artefactual, is bound to be vanishingly small. Ironically, in any combinatorial space, it is the very same restrictive criteria of function which prevent gradual functional change which also isolate all functional systems to vastly improbable and inaccessible regions of the space.

To determine, finally, whether the distribution of islands of function in organic nature conforms to a probable continuum or an improbable discontinuum and to assess definitively the relevance of chance in evolution would be a colossal task. Just as in the case of the sentences and watches, we would have to begin by constructing a multi-dimensional universe filled with all possible combinations of organic chemicals. Within this space of all possibilities there would exist every conceivable functional biological system, including not only those which exist on earth, but all other functional biological systems which could possibly work elsewhere in the universe. The functional systems would range from simple protein molecules capable of particular catalytic functions right up to immensely complex systems such as the human brain. Within this universe of all possibilities we would find many strange biological systems, such as enzymes capable of transforming unique substrates not found on earth, and perhaps nervous systems resembling those found among vertebrates on earth, but far more advanced. We would also find many sorts of complex aggregates, the function of which may not be clear but which we could dimly conceive as being of some value on some alien planet.

Such a space would of course consist mainly of combinations which would have no conceivable biological function – merely junk aggregates ranging from functionless proteins to entirely disordered nervous systems reminiscent of Cuvier's incompatibilities. From the space we would be able to calculate exactly how probable functional biological systems are and how easy it is to go from one functional system to another, Darwinian fashion, in a series of unit mutational steps through functional intermediates. Of course if analogy is any guide then the space would in all probability conjure up a vision of nature more in harmony with the thinking of Cuvier **and early 19th-century typology than modern evolutionary thought in which each island of meaning is intensely isolated unlinked by** transitional forms and quite beyond the reach of chance.

At present we are very far from being able to construct such a space of all organic possibilities and to calculate the probability of functional living systems. Nevertheless, for some of the lower order functional systems, such as individual proteins, their rarity in the space can be at least tentatively assessed.

A protein (as we have seen in Chapter Ten) is fundamentally a long chain-like molecule built up out of twenty different kinds of amino acids. After its assembly the long amino acid chain automatically folds into a specific stable 3D configuration. Particular protein functions depend on highly specific 3D shapes and, in the case of proteins which possess catalytic functions, depend on the protein possessing a particular active site, again of highly specific 3D configuration.

Although the exact degree of isolation and rarity of functional proteins is controversial it is now generally conceded by protein chemists that most functional proteins would be difficult to reach or to interconvert through a series of successive individual amino acid mutations. Zuckerkandl comments:[9]

> Although, abstractly speaking, any polypeptide chain can be transformed into any other by successive amino acid substitutions and other mutational events, in concrete situations the pathways between a poorly and a highly adapted molecule will be mostly impracticable. Any such pathway, whether the theoretically shortest, or whether a longer one, will perforce include stages of favorable change as well as

hurdles. Of the latter some will be surmountable and some not. Some of the latter will presumably be present along the pathways of adaptive change in a very large majority of ill adapted de novo polypeptide chains.

Consequently, when a protein molecule is selected for its weak enzymatic activity and in spite of limited substrate specificity, it will most often represent a dead end road.

The impossibility of gradual functional transformation is virtually self-evident in the case of proteins: mere casual observation reveals that a protein is an interacting whole, the function of every amino acid being more or less (like letters in a sentence or cogwheels in a watch) essential to the function of the entire system. To change, for example, the shape and function of the active site (like changing the verb in a sentence or an important cogwheel in a watch) in isolation would be bound to disrupt all the complex intramolecular bonds throughout the molecule, destabilizing the whole system and rendering it useless. Recent experimental studies of enzyme evolution largely support this view, revealing that proteins are indeed like sentences, and are only capable of undergoing limited degrees of functional change through a succession of individual amino acid replacements. The general consensus of opinion in this field is that significant functional modification of a protein would require several simultaneous amino acid replacements of a relatively improbable nature. The likely impossibility of major functional transformation through individual amino acid steps was raised by Brian Hartley, a specialist in this area, in an article in the journal *Nature* in 1974. From consideration of the atomic structure of a family of closely related proteins which, however, have different amino acid arrangements in the central region of the molecule, he concluded that their functional interconversion would be impossible:[10]

> It is hard to see how these alternative arrangements could have evolved without going through an intermediate that could not fold correctly (i.e. would be non functional).

Here then, is at least one functional subset of the space of all organic possibilities which almost certainly conforms to the general discontinuous pattern observed in the case of other complex systems. But how discontinuous is the pattern of the distribution of proteins within the space of all organic possibilities? Might functional proteins be beyond the reach of chance?

In attempting to answer the question – how rare are functional proteins? – we must first decide what general restrictions must be imposed on a sequence of amino acids before it can form a biologically functional protein. In other words, what are the rules or criteria which govern functionality in an amino acid sequence?

First, a protein must be a stable structure so that it can hold a particular 3D shape for a sufficiently long period to allow it to undergo a specific interaction with some other entity in the cell. Second, a protein must be able to fold into its proper shape. Third, if a protein is to possess catalytic properties it must have an active site which necessitates a highly specific arrangement of atoms in some region of its surface to form this site.

From the tremendous advances that have been made over the past two decades in our knowledge of protein structure and function, there are compelling reasons for believing that these criteria for function would inevitably impose severe limitations on the choice of amino acids. It is very difficult to believe that the criteria for stability and for a folding algorithm would not require a relatively severe restriction of choice in at least twenty per cent of the amino acid chain. To get the precise atomic 3D shape of active sites may well require an absolute restriction in between one and five per cent of the amino acid sequence.

There is a considerable amount of empirical evidence for believing that the criteria for function must be relatively stringent. One line of evidence, for example, is the very strict conservation of overall shape and the exact preservation of the configuration of active sites in homologous proteins such as the cytochromes in very diverse species. Also relevant is the fact that most mutations which cause changes in the amino acid sequence of proteins tend to damage function to a greater or lesser degree. The effects of such mutations have been carefully documented in the case of haemoglobin, and some of them were described in an excellent article in *Nature* by Max Perutz,[11] who himself pioneered the X-ray crystallographic work which first revealed the detailed 3D structure of proteins. As Perutz shows, although many of the amino acids occupying positions on the surface of the molecule can be changed with little effect on function, most of the amino acids in the centre of the protein cannot be changed without having drastic deleterious effects on the stability and function of the molecule.

There are, in fact, both theoretical and empirical grounds for believing that the *a priori* rules which govern function in an amino acid sequence are relatively stringent. If this is the case, and all the evidence points in this direction, it would mean that functional proteins could well be exceedingly rare. The space of all possible amino acid sequences (as with letter sequences) is unimaginably large and consequently sequences which must obey particular restrictions which can be defined, like the rules of grammar, are bound to be fantastically rare. Even short unique sequences just ten amino acids long only occur once by chance in about 10^{13} average-sized proteins; unique sequences twenty amino acids long once in about 10^{26} proteins, and unique sequences thirty amino acids long once in about 10^{39} proteins!

As it can easily be shown that no more than 10^{40} possible proteins could have ever existed on earth since its formation, this means that, if protein functions reside in sequences any less probable than 10^{-40}, it becomes increasingly unlikely that any functional proteins could ever have been discovered by chance on earth.

We have seen in Chapter Eleven that envisaging how a living cell could have gradually evolved through a sequence of simple protocells seems to pose almost insuperable problems. If the estimates above are anywhere near the truth then this would undoubtedly mean that the alternative scenario – the possibility of life arising suddenly on earth by chance – is infinitely small. To get a cell by chance would require at least one hundred functional proteins to appear simultaneously in one place. That is one hundred simultaneous events each of an independent probability which could hardly be more than 10^{-20} giving a maximum combined probability of 10^{-2000}. Recently, Hoyle and Wickramasinghe in *Evolution from Space* provided a similar estimate of the chance of life originating, assuming functional proteins to have a probability of 10^{-20}:[12]

By itself, this small probability could be faced, because one must contemplate not just a single shot at obtaining the enzyme, but a very large number of trials such as are supposed to have occurred in an organic soup early in the history of the Earth. The trouble is that there are about two thousand enzymes, and the chance of obtaining them all in a random trial is only one part in $(10^{20})^{2000} = 10^{40,000}$ an outrageously small probability that could not be faced even if the whole universe consisted of organic soup.

Although at present we still have insufficient knowledge of the rules which govern function in amino acid sequences to calculate with any degree of certainty the actual rarity of functional proteins, it may be that before long quite rigorous estimates may be possible. Over the next few decades advances in molecular biology are inevitably going to reveal in great detail many more of the principles and rules which govern the function and structure of protein molecules. In fact, by the end of the century, molecular engineers may be capable of specifying quite new types of functional proteins. From the first tentative steps in this direction it already seems that, in the design of new functional proteins, chance will play as peripheral a role as it does in any other area of engineering.[13]

The Darwinian claim that all the adaptive design of nature has resulted from a random search, a mechanism unable to find the best solution in a game of checkers, is one of the most daring claims in the history of science. But it is also one of the least substantiated. No evolutionary biologist has ever produced any quantitive proof that the designs of nature are in fact within the reach of chance. There is not the slightest justification for claiming, as did Richard Dawkins recently:[14]

> . . . Charles Darwin showed how it is possible for blind physical forces to mimic the effects of conscious design, and, by operating as a cumulative filter of chance variations, to lead eventually to organised and adaptive complexity, to mosquitoes and mammoths, to humans and therefore, indirectly, to books and computers.

Neither Darwin, Dawkins nor any other biologist has ever calculated the probability of a random search finding in the finite time available the sorts of complex systems which are so ubiquitous in nature. Even today we have no way of rigorously estimating the probability or degree of isolation of even one functional protein. It is surely a little premature to claim that random processes could have assembled mosquitoes and elephants when we still have to determine the actual probability of the discovery by chance of one single functional protein molecule!

NOTES

1. Monod, J. (1972) *Chance and Necessity*, Collins, London, p114.
2. Minsky, M. (1966) "Artificial Intelligence", *Scientific American*, 215(3) September, p246–60, see p247–48.
3. Moorhead, P. S. and Kaplan, M. M., eds (1967) *Mathematical Challenges to the Darwinian Interpretation of Evolution*, Wistar Institute Symposium Monograph.
4. Pattee, H. H. (1966) "Introduction to Session One" in *Natural Automata and Useful Simulations*, eds II. II. Pattee et al, Spartan Books, Washington, pp1 2.
5. Von Neumann, J. (1966) *Theory of Self-Reproducing Automata*, University of Illinois Press, Urbana.
6. Paley, W. (1818) *Natural Theology on Evidence and Attributes of Deity*, 18th ed, Lackington, Allen and Co, and James Sawers, Edinburgh, p13.
7. Jukes, T. H. (1982) "Aircraft Evolution", *Nature*, 295, p548.
8. Simpson, G. G. (1947) "The Problem of Plan and Purpose in Nature", *Scientific Monthly*, 64: 481–495, see p493.
9. Zuckerkandl, E. (1975) "The Appearance of New Structures in Proteins During Evolution", *J. Mol. Evol.*, 7: 1–57, see p21.
10. Rigby, P. W. J., Burleigh, B. D. Jnr, and Hartley, B. S. (1974) "Gene Duplication in Experimental Enzyme Evolution", *Nature*, 251: 200–204, see p200.
11. Perutz, M. F. and Lehmann, H. (1968) "Molecular Pathology of Human Haemoglobin", *Nature*, 219: 902–09.
12. Hoyle, F. and Wickramasinghe, C. (1981) *Evolution from Space*, J. M. Dent and Sons, London, p24.
13. Paba, C. (1983) "Designing Proteins and Peptides", *Nature*, 301:200.
14. Dawkins, R. (1982) "The Necessity of Darwinism", *New Scientist*, 94, (1301) 15 April, pp130–132, see p130.

CHAPTER 14

The Puzzle of Perfection

> Nothing at first can appear more difficult to believe than that the more complex organs and instincts have been perfected, not by means superior to, though analogous with, human reason, but by the accumulation of innumerable slight variations, each good for the individual possessor.

While Darwin was attempting to convince the world of the validity of evolution by natural selection he was admitting privately to friends to moments of doubt over its capacity to generate very complicated adaptations or "organs of extreme perfection", as he described them. In a letter to Asa Gray, the American biologist, written in 1861, just two years after the publication of *The Origin of Species*, he acknowledges these doubts and admits that "The eye to this day gives me a cold shudder."[1]

It is easy to sympathize with Darwin. Such feelings have probably occurred to most biologists at times, for to common sense it does indeed appear absurd to propose that chance could have thrown together devices of such complexity and ingenuity that they appear to represent the very epitome of perfection. There can hardly be a student of human physiology who has not on occasion been struck by the sheer brilliance apparent in the design of so many physiological adaptations. Like, for example, in the elegance manifest in the design of the mammalian kidney which combines so many wonderfully clever adaptations to achieve water and salt homeostasis and the control of blood pressure while at the same time concentrating and eliminating from the body urea, the main end product of nitrogen metabolism. Or like the choice of the bicarbonate buffer system as the body's main defence against the accumulation of metabolic acids. This is a particularly elegant adaptation which exploits the ready availability of bicarbonate base, the main end product of oxidative metabolism, as well as the unique capacity of bicarbonate to combine with hydrogen ions to form water and the innocuous gas carbon dioxide, which can be so conviently eliminated from the body by the lungs to achieve a highly efficient and ingenious system for the maintenance of acid base homeostasis.

Aside from any quantitive considerations, it seems intuitively, impossible that such self-evident brilliance in the execution of design could ever have been the result of chance. For, even if we allow that chance might have occasionally hit on a relatively ingenious adaptive end, it seems inconceivable that it could have reached so many ends of such surpassing "perfection". It is, of course, possible to allude to certain sorts of apparent "imperfections" in life, where an adaptation conveys the impression that nature often makes do in an opportunistic sort of way, moulding the odd lucky accident into something resembling an "imperfect" adaptation. This is the thrust of Gould's argument in his discussion of the curiously elongated bone in the hand of a panda which it uses as a kind of a thumb.[2] Yet, just as a few missing links are not sufficient to close the gaps of nature, a few imperfect adaptations which give every impression of having been achieved by chance are certainly, amid the general perfection of design in nature, an insufficient basis on which to argue for the all-sufficiency of chance. Such imperfections only serve to highlight the fact that, in general, biological adaptations exhibit, as Darwin confessed: "a perfection of structure and coadaptation which justly excites our admiration."[3]

The intuitive feeling that pure chance could never have achieved the degree of complexity and ingenuity so ubiquitous in nature has been a continuing source of scepticism ever since the publication of the *Origin*; and throughout the past century there has always existed a significant minority of first-rate biologists who have never been able to bring themselves to accept the validity of Darwinian claims. In fact, the number of biologists who have expressed some degree of disillusionment is practically endless. When Arthur Koestler organized the Alpbach Symposium in 1969 called "Beyond Reductionism", for the express purpose of bringing together biologists critical of orthodox Darwinism, he was able to include in the list of participants many authorities of world stature, such as Swedish neurobiologist Holgar Hyden, zoologists Paul Weiss and W. H. Thorpe, linguist David McNeil and child psychologist Jean Piaget. Koestler had this to say in his opening remarks:[4] ". . . invitations were confined to personalities in academic life with undisputed authority in their respective fields, who nevertheless share that holy discontent."

At the Wistar Institute Symposium in 1966, which brought together mathematicians and biologists of impeccable academic credentials, Sir Peter Medawar acknowledged in his introductory address the existence of a widespread feeling of scepticism over the role of chance in evolution, a feeling in his own words that:[5] ". . . something is missing from orthodox theory."

Perhaps in no other area of modern biology is the challenge posed by the extreme complexity and ingenuity of biological adaptations more apparent than in the fascinating new molecular world of the cell. Viewed down a light microscope at a magnification of some several hundred times, such as would have been possible in Darwin's time, a living cell is a relatively disappointing spectacle appearing only as an ever-changing and apparently disordered pattern of blobs and particles which, under the influence of unseen turbulent forces, are continually tossed haphazardly in all directions. To grasp the reality of life as it has been revealed by molecular biology, we must magnify a cell a thousand million times until it is twenty kilometres in diameter and resembles a giant airship large enough to cover a great city like London or New York. What we would then see would be an object of unparalleled complexity and adaptive design. On the surface of the cell we would see millions of openings, like the port holes of a vast space ship, opening and closing to allow a continual stream of materials to flow in and out. If we were to enter one of these openings we would find ourselves in a world of supreme technology and bewildering complexity. We would see endless highly organized corridors and conduits branching in every direction away from the perimeter of the cell, some leading to the central memory bank in the nucleus and others to assembly plants and processing units. The nucleus itself would be a vast spherical chamber more than a kilometre in diameter, resembling a geodesic dome inside of which we would see, all neatly stacked together in ordered arrays, the miles of coiled chains of the DNA molecules. A huge range of products and raw materials would shuttle along all the manifold conduits in a highly ordered fashion to and from all the various assembly plants in the outer regions of the cell.

We would wonder at the level of control implicit in the movement of so many objects down so many seemingly endless conduits, all in perfect unison. We would see all around us, in every direction we looked, all sorts of robot-like machines. We would notice that the simplest of the functional components of the cell, the protein molecules, were astonishingly, complex pieces of molecular machinery, each one consisting of about three thousand atoms arranged in highly organized 3-D spatial conformation. We would wonder even more as we watched the strangely purposeful activities of these weird molecular machines, particularly when we realized that, despite all our accumulated knowledge of physics and chemistry, the task of designing one such molecular machine – that is one single functional protein molecule – would be completely beyond our capacity at present and will probably not be achieved until at least the beginning of the next century. Yet the life of the cell depends on the integrated activities of thousands, certainly tens, and probably hundreds of thousands of different protein molecules.

We would see that nearly every feature of our own advanced machines had its analogue in the cell: artificial languages and their decoding systems, memory banks for information storage and retrieval, elegant control systems regulating the automated assembly of parts and components, error fail-safe and proof-reading devices utilized for quality control, assembly processes involving the principle of prefabrication and modular construction. In fact, so deep would be the feeling of *deja-vu*, so persuasive the analogy, that much of the terminology we would use to describe this fascinating molecular reality would be borrowed from the world of late twentieth-century technology.

What we would be witnessing would be an object resembling an immense automated factory, a factory larger than a city and carrying out almost as many unique functions as all the manufacturing activities of man on earth. However, it would be a factory which would have one capacity not equalled in any of our own most advanced machines, for it would be capable of replicating its entire structure within a matter of a few hours. To witness such an act at a magnification of one thousand million times would be an awe-inspiring spectacle.

To gain a more objective grasp of the level of complexity the cell represents, consider the problem of constructing an atomic model. Altogether a typical cell contains about ten million million atoms. Suppose we choose to build an exact replica to a scale one thousand million times that of the cell so that each atom of the model would be the size of a tennis ball. Constructing such a model at the rate of one atom per minute, it would take fifty million years to finish, and the object we would end up with would be the giant factory, described above, some twenty kilometres in diameter, with a volume thousands of times that of the Great Pyramid.

Copying nature, we could speed up the construction of the model by using small molecules such as amino acids and nucleotides rather than individual atoms. Since individual amino acids and nucleotides are made up of between ten and twenty atoms each, this would enable us to finish the project in less than five million years. We could also speed up the project by mass producing those components in the cell which are present in many copies. Perhaps three-quarters of the cell's mass can be accounted for by such components. But even if we could produce these very quickly we would still be faced with manufacturing a quarter of the cell's mass which consists largely of components which only occur once or twice and which would have to be constructed, therefore, on an individual basis. The complexity of the cell, like that of any complex machine, cannot be reduced to any sort of simple pattern, nor can its manufacture be reduced to a simple set of algorithms or programmes. Working continually day and night it would still be difficult to finish the model in the space of one million years.

In terms of complexity, an individual cell is nothing when compared with a system like the mammalian brain. The human brain consists of about ten thousand million nerve cells. Each nerve cell puts out somewhere in the region of between ten thousand and one hundred thousand connecting fibres by which it makes contact with other nerve cells in the brain. Altogether the total number of connections in the human brain approaches 10^{15} or a thousand million million. Numbers in the order of 10^{15} are of course completely beyond comprehension. Imagine an area about half the size of the USA (one million square miles) covered in a forest of trees containing ten thousand trees per square mile. If each tree contained one hundred thousand leaves the total number of leaves in the forest would be 10^{15}, equivalent to the number of connections in the human brain!

Despite the enormity of the number of connections, the ramifying forest of fibres is not a chaotic random tangle but a highly organized network in which a high proportion of the fibres are unique adaptive communication channels following their own specially ordained pathway through the brain. Even if only one hundredth of the connections in the brain were specifically organized, this would still represent a system containing a much greater number of specific connections than in the entire communications network on Earth. Because of the vast number of unique adaptive connections to assemble an object remotely resembling the brain, would take an eternity even applying the most sophisticated engineering techniques.

Undoubtedly, the complexity of biological systems in terms of the sheer number of unique components is very impressive; and it raises the obvious question: could any sort of purely random process ever have assembled such systems in the time available? As all the complexity of a living system is reducible ultimately to its genetic blueprint, the really crucial question to ask is what is the sum total of all the unique adaptive genetic traits necessary for the specification of a higher organism like a mammal? In effect, how many genes are there in the genomes of higher organisms? And how many unique adaptive features are there in each individual gene?

We have seen in Chapter Ten that each gene is a sequence of DNA about one thousand nucleotides long. If only ten percent of the nucleotides are adaptively critical for the specification of the encoded protein then each gene would contain one hundred unique adaptive traits or significant bits of information. This is likely to be a minimum estimate because as we have also seen in Chapter Ten the whole question of gene number and complexity has been revolutionized recently by the discovery that most genes in higher organisms are split. It is now clear that the process of gene expression is far more complex than seemed possible only a few years ago and that a considerable number of unique cutting and re-splicing events are necessary for the assembly of each different mRNA molecule in all higher organisms. As it seems likely that much of the information which orders these precise recombinational events resides in the actual gene sequence itself, then the number of significant bits of information in most genes probably varies between one hundred and one thousand.

The really significant aspect of the split-gene phenomenon in this context is not so much that it greatly complicates each individual gene, but rather that it provides a recombinational mechanism for greatly expanding the total number of genes in the genomes of higher organisms. As it is, even without any sort of recombinational expansion, there is sufficient DNA in higher organisms to specify for more than one million genes. With so much DNA it is obvious that, by exploiting recombinational possibilities, the total number of genes could be expanded to a figure far in excess of one million.

Is it possible that the gene number in higher organisms might be expanded in this way? There is already a growing, and increasingly irresistible body of evidence pointing in this direction. The mechanism is already known to occur in the immune system and in the genomes of DNA viruses, which closely resemble the genomes of higher organisms. Further, there exists the possibility, for which there is already some suggestive evidence, that the development of specific connections in the brain may necessitate the tagging of individual nerve cells or small subsets of neurones with specific biochemical markers; and this in itself might call for as many as ten thousand million genes as this is the number of nerve cells in the mammalian brain.

If it does turn out over the next few years that this recombinational mechanism is being used to achieve a vast expansion in the total number of genes in higher organisms, then it could well be that the total number of unique adaptive traits in, say, mammalian genomes is in the order of 10^{13} (10^{10} genes, each containing 10^3 significant bits of information). Which could pose what would seem to be an almost insurmountable "numbers problem" for Darwinian theory – a problem of such dimension that it would render all other anti-Darwinian arguments superfluous.

But it is not just the complexity of living systems which is so profoundly challenging, there is also the incredible ingenuity that is so often manifest in their design. Ingenuity in biological design is particularly striking when it is manifest in solutions to problems analogous to those met in our own technology. Without the existence of the camera and the telescope, much of the ingenuity in the design of the eye would not have been perceived. Although the anatomical components of the eye were well known by scientists in the fifteenth century, the ingenuity of its design was not appreciated until the seventeenth century when the basic optics of image formation were first clearly expressed by Kepler and later by Descartes. However, it was only in the eighteenth and nineteenth centuries, as the construction of optical instruments became more complicated, utilizing a movable iris, a focusing device, and corrections for spherical and chromatic aberration, all features which have their analogue in the eye, that the ingenuity of the optical system could at last be appreciated fully by Darwin and his contemporaries.

We now know the eye to be a far more sophisticated instrument than it appeared a hundred years ago. Electro-physiological studies have recently revealed very intricate connections among the nerve cells of the retina, which enable the eye to carry out many types of preliminary data processing of visual information before transmitting it in binary form to the brain. The cleverness of these mechanisms has again been underlined by their close analogy to the sorts of image intensification and clarification processes carried out today by computers, such as those used by NASA, on images transmitted from space. Today it would be more accurate to think of a television camera if we are looking for an analogy to the eye.

There are dozens of examples where advances in technology have emphasized the ingenuity of biological design. One fascinating example of this was the construction of the Soviet lunar exploratory machine, the *Lunakod*, which moved by articulated legs. Legs, rather than wheels, were chosen because of the much greater ease with which an articulated machine could traverse the uneven terrain likely to be met on the lunar surface. Altogether, the *Lunakod* eerily resembled a giant ant, so much so that it was no longer possible to look on the articulated legs of an insect without a new sense of awe and the realization that what one had once taken for granted, and superficially considered a simple adaptation, represented a very sophisticated technological solution to the problem of mobility over an uneven terrain. The control mechanisms necessary to coordinate the motion of articulated legs are far more complicated than might be imagined at first sight. As Raibert and Sutherland, who are currently working in this area, admit.[6]

> It is clear that very sophisticated computer-control programs will be an important component of machines that smoothly crawl, walk or run.

But it is at a molecular level where the analogy between the mechanical and biological worlds is so striking, that the genius of biological design and the perfection of the goals achieved are most pronounced. Take, for example, the problem of information storage, various solutions of which have been utilized in human societies: for thousands of years, information has been stored in written symbols on clay tablets, paper scrolls, and in books. But nowadays the acquisition of information is accelerating so quickly that the printed page is rapidly becoming obsolete and more economical and sophisticated means of storing that information will soon be essential. Already information is being stored on microfilm. However, ultimately even microfilm will become too inefficient and we may be forced to start developing ways of storing information in chemical codes, which would reduce a text book to a microscopic dot. The problems involved in developing chemical coding devices are currently being considered, but to date no one has been able to work out a practical solution.

A chemical solution to the problem of information storage has, of course, been solved in living things by exploiting the properties of the long chain-like DNA polymers in which cells store their hereditary information. It is a superbly economical solution. The capacity of DNA to store information vastly exceeds that of any other known system; it is so efficient that all the information needed to specify an organism as complex as man weighs less than a few thousand millionths of a gram. The information necessary to specify the design of all the species of organisms which have ever existed on the planet, a number according to G. G. Simpson of approximately one thousand million,[7] could be held in a teaspoon and there would still be room left for all the information in every book ever written.

The genius of biological design is also seen in the cell's capacity to synthesize organic compounds. Living things are capable of synthesizing exactly the same sorts of organic compounds as those synthesized by organic chemists. Each of the chemical operations necessary to construct a particular compound is carried out by a specific molecular machine known as an enzyme. Each enzyme is a single large protein molecule consisting of some several thousand atoms linked together to form a particular spatial configuration which confers upon the molecule the capacity to carry out a unique chemical operation. When a number of enzymes are necessary for the assembly of a particular compound, they are arranged adjacent to each other so that, after each step in the operation, the partially completed compound can be conveniently passed to the next enzyme which performs the next chemical operation and so on until the compound is finally assembled. The process is so efficient that some compounds can be assembled in less than a second, while in many cases the same synthetic operations carried out by chemists, even in a well-equipped lab, would take several hours or days or even weeks.

Automated assembly is another feature which has reached its epitome in living systems. Except for relatively simple pieces of machinery – parts of television sets, ball bearings, milk bottles – fully automated production has not yet been achieved in our technology. The cell, however, manufactures all its component structures, even the most complex, by fully automated assembly techniques which are perfectly regulated and controlled. Unlike our own pseudo-automated assembly plants, where external controls are being continually applied, the cell's manufacturing capacity is entirely self-regulated.

Modern technology is constantly striving for increased levels of miniaturization. Consider the *Viking* biology laboratory which recently landed on Mars. Although only one cubic foot in volume it could carry out as many chemical operations as a university laboratory, and involved some forty thousand functional components – a genuinely incredible achievement! However, as we have seen, every living cell is a veritable automated factory depending on the functioning of up to one hundred thousand unique proteins each of which can be considered to be a basic working component analogous to one of the components in the *Viking* lab. Each protein is itself a very complex object, a machine very much more sophisticated than any of the componets of the *Viking* biology lab, consisting of several thousand atoms, all of which are specifically orientated in space. For the purpose of this comparison, we will ignore the extra complexity of each of the cell's working components. A typical cell might have a diameter of 20μ and a volume of roughly $4000c\mu$: the volume of the biology lab on the *Viking* space craft was one cubic foot, or approximately $10^{16}c\mu$, some 10^{13} times greater than the volume of a living cell containing an equivalent number of components. This comparison does not detract from the genius of our technology; it merely emphasizes the quite fantastic character of the technology realized in living systems.

In the near future, one of the major technological challenges facing our culture will be the development of a new source of energy. A solution to the problem of extracting solar energy was solved three and a half thousand million years ago when life began on Earth. The solution is the chloroplast, which is a micro-miniature solar energy plant which converts the light of the sun into sugar – the hydrocarbon fuel which ultimately energizes every cell on Earth. It was also the chloroplast that was the original source of all the fossil fuels upon which our technology is so crucially dependent, and without which the process of industrialization could never have begun.

Everyone today is familiar with artificial languages, such as those used in computers where information is stored in coded form in long linear sequences. Precisely the same technique is utilized by living systems. In all human languages and in all artificial coding systems, individual messages are encoded in discrete successive sequences. Sentences, for example, never overlap. After the elucidation of the genetic code and the realization that genetic information was stored in DNA in a way analogous to that of other coding systems, it was assumed almost universally that genes, like sentences, would be discrete non-overlapping sequences, each restricted to a particular linear region in the DNA.

However, a few years ago a surprising discovery was made by a group of biochemists at Cambridge University. While working on the DNA of a small virus, they discovered that it contained more information than could be accounted for if the genes were arranged in a linear array of discrete sequences. For some time this discrepancy was very puzzling and the explanation, when it came, astonished the biological world. After the exact sequence of all the DNA of the virus had been worked out, the discovery was made that in certain regions two genes were embedded together in the same sequence, that is to say they overlapped.[8]

When two genes overlap in the same sequence the information for both encoded proteins is contained in one DNA sequence in the same way as one sequence of symbols in morse code can contain information for two words and be read in two different ways:

Thus the discrepancy between the coding potential and the number of proteins synthesized was explained by a mechanism of wonderful ingenuity.

Overlapping genes are not the only recently discovered ingenious device for compacting information with great economy into DNA sequences. DNA does not consist entirely of genes containing encoded messages for the specification of proteins; a considerable proportion is involved in control purposes, switching off and on different genes at different times and in different cells. This was considered, again by analogy with human information retrieval systems such as might be used in a library or filing system or computer, to be positioned adjacent to, but separate from, the genes under its control. There was some empirical support for this very logical view but, once more, as in the case of overlapping genes, biological design turned out to be far more clever than was suspected, for it has now been found that many sequences of DNA which perform the crucial control functions related to information retrieval are situated not adjacent to the genes which they control but actually embedded within the genes themselves.

Another compacting device, which has been shown to be utilized in living systems and which again has no strict analogy in our own technology, is the use of the breakdown products of proteins to perform all sorts of functions often quite unrelated to the original function of the "mother" protein. Thus, many protein functions are compacted into an original molecule. The process begins by the synthesis of the original protein which, after performing its function, is broken down in the cell into two smaller proteins, each of which perform two further functions. These two proteins are again broken down into still smaller proteins capable of yet further functions. The device is somewhat analogous to having a whole tool kit compacted within the first tool we require to initiate a particular operation; and when the initial operation is complete, the tool breaks down into the next two tools required for the operation, and so on until the operation is complete.

One of the accomplishments of living systems which is, of course, quite without any analogy in the field of our own technology is their capacity for self-duplication. With the dawn of the age of computers and automation after the Second World War, the theoretical possibility of constructing self-replicating automata was considered seriously by mathematicians and engineers. Von Neumann discussed the problem at great length in his famous book *Theory of Self-Reproducing Automata*,[9] but the practical difficulties of converting the dream into reality have proved too daunting. As Von Neumann pointed out, the construction of any sort of self-replicating automaton would necessitate the solution to three fundamental problems: that of storing information; that of duplicating information; and that of designing an automatic factory which could be programmed from the information store to construct all the other components of the machine as well as duplicating itself. The solution to all three problems is found in living things and their elucidation has been one of the triumphs of modern biology.

So efficient is the mechanism of information storage and so elegant the mechanism of duplication of this remarkable molecule that it is hard to escape the feeling that the DNA molecule may be the one and only perfect solution to the twin problems of information storage and duplication for self-replicating automata.

The solution to the problem of the automatic factory lies in the ribosome. Basically, the ribosome is a collection of some fifty or so large molecules, mainly proteins, which fit tightly together. Altogether the ribosome consists of a highly organized structure of more than one million atoms which can synthesise any protein that it is instructed to make by the DNA, including the particular proteins which compromise its own structure – so the ribosome can construct itself!

The protein synthetic apparatus is also, however, the solution to an even deeper problem than that of self-replication. Proteins can be designed to perform structural, logical, and catalytic functions. For instance, they form the impervious materials of the skin, the contractile elements of muscles, the transparent substance of the lens of the eye: and, because of their practically unlimited potential, almost any conceivable biochemical object can be ultimately constructed using these remarkable molecules as basic structural and functional units. The choice of the protein synthetic apparatus as the solution to the problem of the automatic factory has deep implications. Not only does it represent a solution to one of the problems of designing a self-duplicating machine but it also represents a solution to an even deeper problem, that of constructing a universal automaton. The protein synthetic apparatus cannot only replicate itself but, in addition, if given the correct information, it can also construct any other biochemical machine, however great its complexity, just so long as its basic functional units are comprised of proteins, which, because of the near infinite number of uses to which they can be put, gives it almost limitless potential.

It is astonishing to think that this remarkable piece of machinery, which possesses the ultimate capacity to construct every living thing that ever existed on Earth, from a giant redwood to the human brain, can construct all its own components in a matter of minutes and weigh less than 10^{-16} grams. It is of the order of several thousand million million times smaller than the smallest piece of functional machinery ever constructed by man.

Human intelligence is yet another achievement of life which has not been equalled in our technology, despite the tremendous effort and some significant advances which have been made in the past two decades towards the goal of artificial intelligence – a goal which may still be further away than is often assumed. As David Waltz points out in a recent article in the *Scientific American*, no machines have yet been constructed which can in any significant way mimic the cognitive capacities of the human brain. The most telling criticism of current work in artificial intelligence is that it has not been successful in modelling what is called common sense. As Waltz explains, we still do not understand how the human brain thinks:[10]

> substantially better models of human cognition must be developed before systems can be designed that will carry out even simplified versions of common-sense tasks. I expect the development of such models to keep me and many others fascinated for a long time.

It could turn out that both self-duplication and intelligence cannot be achieved in terms of a non-biological plastics' and metals' technology. Perhaps a fully intelligent machine, ie one that could mimic the intelligence of man, requires a structure approaching the complexity of the human brain which could mean, as we have seen above, that the goal may never be reached, for an object of this complexity would require eternity for its assembly in terms of our current engineering capabilities.

The eerie artefact-like character of life and the analogy with our own advanced machines has an important philosophical consequence, for it provides the means for a powerful reformulation of the old analogical argument to design which has been one of the basic creationist arguments used throughout western history – going back to Aristotle and presented in its classic form by William Paley in his famous watch-to-watchmaker discourse.

According to Paley,[11] we would never infer in the case of a machine, such as a watch, that its design was due to natural processes such as the wind and rain; rather, we would be obliged to postulate a watchmaker. Living things are similar to machines, exhibiting the same sort of adaptive complexity and we must, therefore, infer by analogy that their design is also the result of intelligent activity.

One of the principal weaknesses of this argument was raised by David Hume,[12] who pointed out that organisms may be only superficially like machines but natural in essence. Only if an object is strikingly analogous to a machine in a very profound sense would the inference to design be valid. Hume's criticism is generally considered to have fatally weakened the basic analogical assumption upon which the inference to design is based, and it is certainly true that neither in the eighteenth century nor at any time during the past two centuries has there been sufficient evidence for believing that living organisms were like machines in any profound sense.

It is only possible to view an unknown object as an artefact if its design exploits well-understood technological principles and its creation can be precisely envisaged. For this reason, stone age man would have had great difficulty in recognizing the products of twentieth-century technology as machines and we ourselves would probably experience the same bewilderment at the artefacts of a technological civilization far in advance of our own.

How would stone age man have judged a motor car or a pocket calculator? Incapable of manufacturing anything other than a crudely shaped flint tool, so primitive that it could hardly be distinguished from a natural piece of rock, the inside of a pocket calculator would seem a purposeless tangle of strings – a random maze of straw trapped inside a leather bag. Even megalithic monuments like Stonehenge or the Pyramids, artefacts which are primitive from our twentieth century standpoint, would cause considerable confusion to a paleolithic man. How would an ancient Egyptian have judged an airplane or a submarine? Only if our ancestors had seen a man in the cockpit of the airplane would they have grasped the incredible, that it was an artefact. It would, of course, be an artefact beyond their comprehension – an artefact of the gods.

It has only been over the past twenty years with the molecular biological revolution and with the advances in cybernetic and computer technology that Hume's criticism has been finally invalidated and the analogy between organisms and machines has at last become convincing. In opening up this extraordinary new world of living technology biochemists have become fellow travellers with science fiction writers, explorers in a world of ultimate technology, wondering increduously as new miracles of atomic engineering are continually brought to light in the course of their strange adventure into the microcosm of life. In every direction the biochemist gazes, as he journeys through this weird molecular labyrinth, he sees devices and appliances reminiscent of our own twentieth-century world of advanced technology. In the atomic fabric of life we have found a reflection of our own technology. We have seen a world as artificial as our own and as familiar as if we had held up a mirror to our own machines.

Paley was not only right in asserting the existence of an analogy between life and machines, but was also remarkably prophetic in guessing that the technological ingenuity realized in living systems is vastly in excess of anything yet accomplished by man.[13]

> Every indication of contrivance, every manifestation of design which existed in the watch exists in the works of nature with the difference, on the side of nature, being greater and more, and that in a degree which exceeds all computation . . . yet in a multitude of cases, are not less evidently mechanical, not less evidently contrivances, . . . than are the most perfect productions of human ingenuity.

The almost irresistible force of the analogy has completely undermined the complacent assumption, prevalent in biological circles over most of the past century, that the design hypothesis can be excluded on the grounds that the notion is fundamentally a metaphysical *a priori* concept and therefore scientifically unsound. On the contrary, the inference to design is a purely *a posteriori* induction based on a ruthlessly consistent application of the logic of analogy. The conclusion may have religious implications, but it does not depend on religious presuppositions.

If we are to assume that living things are machines for the purposes of description, research and analysis, and for the purposes of rational and objective debate, as argued by Michael Polyani[14] and Monod[15] among many others, there can be nothing logically inconsistent, as Paley would have argued, in extending the usefulness of the analogy to include an explanation for their origin.

It is interesting to speculate how the theory of natural selection might have fared in the nineteenth century had the analogy between the living and mechanical worlds been as apparent then as it is today. The depth of the machine-organism analogy would have more than satisfied William Paley, and would certainly have provided Darwin's antagonists with powerful ammunition with which to resist the idea of natural selection.

Although the argument for design has been unfashionable in biology for the past century, the feeling that chance is an insufficient means of achieving complex adaptations has continually been expressed by a dissenting minority, and this dissent is undiminished today. As we have seen, the dissenters have not only been drawn from the ranks of fundamentalists, Lamarckists and vitalists such as Bergson and

Teilhard de Chardin, but also from very respectable members of the scientific establishment.

It is the sheer universality of perfection, the fact that everywhere we look, to whatever depth we look, we find an elegance and ingenuity of an absolutely transcending quality, which so mitigates against the idea of chance. Is it really credible that random processes could have constructed a reality, the smallest element of which – a functional protein or gene – is complex beyond our own creative capacities, a reality which is the very antithesis of chance, which excels in every sense anything produced by the intelligence of man? Alongside the level of ingenuity and complexity exhibited by the molecular machinery of life, even our most advanced artefacts appear clumsy. We feel humbled, as neolithic man would in the presence of twentieth-century technolgy.

It would be an illusion to think that what we are aware of at present is any more than a fraction of the full extent of biological design. In practically every field of fundamental biological research ever-increasing levels of design and complexity are being revealed at an ever-accelerating rate. The credibility of natural selection is weakened, therefore, not only by the perfection we have already glimpsed but by the expectation of further as yet undreamt of depths of ingenuity and complexity. To those who still dogmatically advocate that all this new reality is the result of pure chance one can only reply, like Alice, incredulous in the face of the contradictory logic of the Red Queen:[16]

> Alice laughed. "There's no use trying", she said. "One can't believe impossible things". "I dare say you haven't had much practice," said the queen. "When I was your age I did it for half an hour a day. Why sometimes I've believed as many as six impossible things before breakfast."

NOTES

1. Darwin, C. (1860) in letter to Asa Gray in *Life and Letters of Charles Darwin* (1888) 3 vols, ed F. Darwin, John Murray, London, vol 2, p 273.
2. Gould, S. J. (1980) *The Panda's Thumb*, W. W. Norton and Co, Inc, New York and London, see Chapter One.
3. Darwin, C. (1860) *The Origin of Species*, 6th ed (1962) Collier Books, New York, p 26.
4. Koestler, A. (1969) *Beyond Reductionism*, Hutchinson & Co Ltd, London, p 2.
5. Medawar, P. (1966) Remarks by chairman in *Mathematical Challenges to the*

Darwinian Interpretation of Evolution, Wistar Institute Symposium Monograph, vol 5 xi.

6. Raibert, M. H. and Sutherland, I. E. (1983) "Machines that Walk", *Scientific American*, 248(1): 32–41, p 32.

7. Simpson, G. G. (1960) "The History of Life" in *Evolution of Life*, ed Sol Tax, University of Chicago Press, Chicago, pp 117–180, see p 135.

8. Barrell, B. G., Air, G. M. and Hutchinson, C. A. III (1976) "Overlapping Genes in Bacteriophage ΘX174", *Nature* 264: 34–41.

9. Von Neumann, J. (1966) *Theory of Self-Reproducing Automata*, University of Illinois Press, Urbana.

10. Waltz, D. L. (1982) "Artificial Intelligence", *Scientific American*, 247(4), pp 101–122.

11. Paley, W. (1818) *Natural Theology on Evidences and Attributes of the Deity*, 18th ed, Lackington, Allen and Co, and James Sawers, Edinburgh, Chapter One.

12. Hume, D. (1779) *Dialogues Concerning Natural Religion*, Fontana Library Edition (1963), Collins, London, part 7, p 149.

13. Paley, op cit, p 22.

14. Polanyi, M. (1968) "Life's Irreducible Structure", *Science*, 160: 1308–12.

15. Monod, J. (1972) Chance and Necessity, Collins, London.

16. Carroll, L. (1880) *Alice through the Looking-Glass*, Macmillan and Co, London, p 100.

CHAPTER 15

The Priority of the Paradigm

When on board H.M.S. 'Beagle', as naturalist, I was much struck with certain facts in the distribution of the organic beings inhabiting South America. . . . These facts, as will be seen in the latter chapters of this volume, seemed to throw some light on the origin of species – that mystery of mysteries, as it has been called by one of our greatest philosophers.

Since 1859, a vast amount of evidence has accumulated which has thoroughly substantiated Darwin's views as far as microevolutionary phenomena are concerned. Evolution by natural selection has been directly observed in nature, and it is beyond any reasonable doubt that new reproductively isolated populations – species – do in fact arise from pre-existing species. Although some of the details of the process are still controversial, and certain aspects of the modern view of speciation differ slightly from Darwin's, it is clear that the process involves a gradual accumulation of small genetic changes guided mainly by natural selection.

But while his special theory has been confirmed, its general application, the grand claim that, in Mayr's words:[1]

. . . all evolution is due to the accumulation of small genetic changes guided by natural selection and that transpecific evolution is nothing but an extrapolation and magnification of the events which take place within population and species . . .

remains as unsubstantiated as it was one hundred and twenty years ago. The very success of the Darwinian model at a microevolutionary level, and particularly the mode of its success – by rigorous empirical documentation of actual evolutionary events and thoroughly worked out models showing precisely how the process of speciation and microevolution occurs – only serves to highlight its failure at a macroevolutionary level.

Neither of the two fundamental axioms of Darwin's macroevolutionary theory – the concept of the continuity of nature, that is the idea of a functional continuum of all life forms linking all species together and ultimately leading back to a primeval cell, and the belief that all the adaptive design of life has resulted from a blind random process – have been validated by one single empirical discovery or scientific advance since 1859. Despite more than a century of intensive effort on the part of evolutionary biologists, the major objections raised by Darwins's critics such as Agassiz, Pictet, Bronn and Richard Owen have not been met. The mind must still fill up the "large blanks" that Darwin acknowledged in his letter to Asa Gray.

One hundred and twenty years ago it was possible for a sceptic to be forgiving, to give Darwinism the benefit of the doubt and to allow that perhaps future discoveries would eventually fill in the blanks that were so apparent in 1859. Such a position is far less tenable today.

Since the birth of modern biology in the mid-eighteenth century, nearly all advocates of the continuity of nature have attempted to explain away the gaps in terms of what ultimately amounts to some sort of sampling error hypothesis. Very few professional biologists have adopted the alternative nominalist position and explained them away as convenient and arbitrary inventions of the mind.

That the gaps cannot be dismissed as inventions of the human mind, merely figments of an anti-evolutionary imagination – an imagination prejudiced by typology, essentialism or creationism – is amply testified by the fact that their existence has always been just as firmly acknowledged by the advocates of evolution and continuity. While it may have been the anti-evolutionists who, in perceiving the enormity of the empirical challenge posed by the existence of breaks in the order of nature, coined the phrase "missing links", it has been the evolutionists who have acknowledged their existence, who have sought them with such persistence.

As firm believers in the continuity of organic nature, the eighteenth century rationalists who adhered to the doctrine of the great chain of being were no less excited by the discovery of "missing links" than were their evolutionary fellow travellers after 1859. As Lovejoy comments:[2]

> . . . it was in the eyes of the eighteenth century, a great moment in the history of science when Trembley in 1739 rediscovered the fresh-water polyp Hydra* (it had already been observed by Leeuwenhoek), this creature being at once hailed as the long-sought missing link between plants and animals. This and similar discoveries in turn served to strengthen the faith in continuity as an a priori rational law of nature . . .

Although the sampling error hypothesis is of course basically a tautology, being derived entirely from an *a priori* belief in the continuity of nature and invented primarily to justify that belief in the face of the self-evident discontinuous appearance of nature, this does not mean that it is wrong or ineffective. However, its validity is open to a very simple test. If the gaps really are due to a sampling error, then by increasing the scope and intensity of the sampling we should see the gaps inexorably narrow, first to reveal a clear, if still broken, sequential pattern, and finally to reveal a perfect continuum of forms linking all known forms together.

That the credibility of the sampling error hypothesis depends on the satisfaction of the crucial condition, that the gaps inexorably narrow as the intensity and scope of the sampling is increased, was apparent in the eighteenth century no less than it was to Darwin. As Bonnet remarked two centuries ago:[3]

> . . . Nature seems to make a great leap in passing from vegetable to the fossil (i.e., rock): there are no bonds, no links known to us, which unite the vegetable and the mineral kingdoms. But shall we judge of the chain of beings by our present knowledge? Because we discover some interruptions, some gaps in it here and there, shall we conclude that these gaps are real? . . . The gap that we find between the vegetable and the mineral will apparently some day be filled up. There was a similar gap between the animal and vegetable; the polyp has come to fill it and to demonstrate the admirable gradation there is between all beings.

It would require a highly prejudicial reading of biological history to conclude that advances in biological knowledge have continually tended to narrow the gaps. On the contrary, the gaps are as intense today as they were in the days of Linnaeus, and almost every major advance in biological knowledge, from the founding of comparative

---

*The hydra, like so many "links", eventually turned out to belong to a clearly defined group which was in no way intermediate in an evolutionary sense. The hydra is now known to belong to the animal phylum Coelenterata.

anatomy and paleontology in the eighteenth century to the recent discoveries of molecular biology, has only tended to emphasize the depth and profundity of the great divisions of nature. Admittedly, the idea that advances in biological knowledge have tended on the whole to intensify the divisions runs counter to the whole ethos of evolutionary thought, yet in certain areas this is so evidently true that not even the most committed evolutionist could pretend that at least in some cases the discontinuities have vastly increased.

The classic example of this, of a major discontinuity being enhanced rather than diminished by advances in knowledge, is the division between life and inorganic nature. In the mid-nineteenth century and perhaps even as late as the 1940s, it was perfectly reasonable to suppose that there was no absolute break, that there was possibly a continuum of simple replicating systems leading from chemistry to life. We now know, as a result of discoveries made over the past thirty years, that not only is there a distinct break between the animate and inanimate worlds but that it is one of the most dramatic in all nature, absolutely unbridged by any series of transitional forms and like so many of the other major gaps of nature, the transitional forms are not only empirically absent but are also conceptually impossible.

The fact that the gaps have not been narrowed in any significant sense since the mid-eighteenth century means that the sampling error strategem has essentially failed, and its failure has in effect stripped the Darwinian concept of a continuum of functional forms leading gradually across all the divisions of nature of any objective basis. Darwin's prediction and hope expressed in the *Origin*, that future discoveries would fill in the blanks, has, on any unprejudiced reading of the evidence, not been fulfilled. On the contrary, the suspicion of his critics, such as Pictet and Owen, that the divisions were fundamental and would never be closed by further sampling, has been confirmed.

Similarly, the credibility of the second great axiom of the Darwinian world view, the all-sufficiency of pure chance as the creative agency of evolution, is greatly diminished since Darwin's day.

There are only two ways to justify the idea of chance as the author of biological design: to calculate the probability of the discovery by chance of functional organic systems, or to test the creative efficiency of random searches in systems which are in every way analogous to living organisms.

Although at present there is still no way of estimating rigorously the probability of a random search discovering functional organic systems, it is abundantly clear that in every analogous system, pure unguided random events cannot achieve any sort of interesting or complex end. As the analogy deepens between organism and machine, as life at a molecular level takes on increasingly the appearance of a sophisticated technology and living organisms the appearance of advanced machines, then the failure to simulate Darwinian evolution in artificial systems increasingly approaches a formal logical disproof of Darwinian claims.

Yet no matter how convincing such disproofs might appear, no matter how contradictory and unreal much of the Darwinian framework might now seem to anyone not committed to its defence, as philosophers of science like Thomas Kuhn[4] and Paul Feyerabend[5] have pointed out, it is impossible to falsify theories by reference to the facts or indeed by any sort of rational or empirical argument. The history of science amply testifies to what Kuhn[6] has termed the "priority of the paradigm" and provides many fascinating examples of the extraordinary lengths to which members of the scientific community will go to defend a theory just as long as it holds sufficient intrinsic appeal.

The defence by medieval astronomers of the Ptolemaic theory of the heavens, and by the eighteenth-century chemists of the phlogiston theory of combustion, provide classic examples. The geocentric theory was the established theory of astronomy from the close of the classical era until its replacement by the Copernican or heliocentric model, a process which was only completed in the early decades of the seventeenth century. Although the geocentric theory was not the only theory proposed to account for the movements of the heavenly bodies – the heliocentric alternative was considered by a variety of Greek astronomers – by the late middle ages it had become a self-evident truth, the one and only sacred and unalterable picture of cosmological reality.

As with every erroneous theory there were certain facts which could not be adequately explained. One particular feature of the movement of the planets had always been a puzzle to astronomers – that of retrograde movement. Mars, for example, appears to move first in one direction across the sky and then stops and starts moving back in the opposite direction. After a short time, it appears to stop and yet again reverse direction so that it finally appears to regain its initial trajectory across the sky. To account for this, the geocentric astronomers invented the following model. They supposed that Mars moved round a circular orbit called an epicycle, the centre of which was moving itself round another circular orbit, a deferent which was centred on the Earth. The motion generated by movement along the deferent and coincidental movement around the epicycle accounted for the curious retrograde motion of the planet. The concept of the epicycle was one of the characteristic features of the Ptolemaic system.

In the centuries following its formulation, the gradual accumulation of astronomical data by medieval Christian and Moslem astronomers revealed further irregularities in the movements of the planets which required further adjustments to the traditional geocentric system. To account for these irregularities, more and more epicycles were proposed and as time went on the theory underwent successive modifications and amendments. By the early sixteenth century the whole Ptolemaic system had become, in the words of a contemporary astronomer, "a monstrosity", a fantastically involved system entailing a vast and ever-growing complexity of epicycles. The state of astronomy is described in Kuhn's *The Structure of Scientific Revolutions*:[7]

> By the thirteenth century Alfonso X could proclaim that if God had consulted him when creating the universe, he would have received good advice. In the sixteenth century, Copernicus' co-worker, Domenico da Novara, held that no system so cumbersome and inaccurate as the Ptolemaic had become could possibly be true of nature. And Copernicus himself wrote in the Preface to the *De Revolutionibus* that the astronomical tradition he inherited had finally created a monster.

However, so ingrained was the idea that the Earth was the centre of the universe that hardly anyone, even those astronomers who were well aware of the growing unreality of the whole system, ever bothered to consider an alternative theory.

The basic underlying concept of phlogiston chemistry was the idea that substances lost something on combustion. Phlogiston was the supposed matter and principle of fire, but not fire itself. The phlogiston theory assumed that all combustible bodies, including metals, contained a common material, phlogiston, which escaped on combustion but could be readily transferred from one body to another. The phlogiston could be restored to any burnt substance or metallic ash by heating them with substances rich in phlogiston, such as charcoal or oil. Hence zinc on heating to redness burns with a brilliant flame as phlogiston supposedly escapes. The resulting white residue, known as the calx of zinc, was metallic zinc minus phlogiston. Therefore, by phlogiston theory: zinc = calx of zinc + phlogiston.

If the calx of zinc was heated with a compound rich in phlogiston then some of the phlogiston was transferred to the calx of zinc and zinc was reformed. Similarly, if phosphorus was burned it produced an acid substance; while it burned much heat and light were generated as the phlogiston escapes. Therefore, by the phlogiston theory: phosphorus = acid + phlogiston.

The theory of phlogiston was an inversion of the true nature of combustion. Removing phlogiston was in reality adding oxygen, while adding phlogiston was actually removing oxygen. The theory was a total misrepresentation of reality. Phlogiston did not even exist, and yet its existence was firmly believed and the theory adhered to rigidly for nearly one hundred years throughout the eighteenth century.

Throughout the phlogiston period certain facts were known which were difficult to reconcile with the theory. For example, it had been known since as early as the sixteenth century that metals increased in weight on combustion, while phlogiston was supposed to have escaped! This awkward fact was either disregarded or was explained away by the implausible strategem of assuming phlogiston to have negative weight. Others concluded that the weight increase occurred by the addition of certain foreign particles mixed with the air and separated by the action of heat. It was all very confusing.

Again, as in the case of the geocentric theory, as time went on discoveries were made which were increasingly difficult to fit into the phlogiston theory, and the theory was modified by the insertion of more and more unwarranted and *ad hoc* assumptions about the nature of phlogiston.

It was discovered, for example, by the Swedish chemist Scheele that when combustion occurred in a confined space, for instance in a bell jar over water, there was a decrease in the volume of air by exactly one quarter. This created a paradox. Firstly, if phlogiston was given off why did the remaining air weigh less and, secondly, why did the remaining air extinguish a taper if the phlogiston has left the object being burnt and entered the atmosphere? According to phlogiston theory the atmosphere should be rich in phlogiston and able to support active combustion.

As experimentation continued the properties of phlogiston became more bizarre and contradictory. But instead of questioning the existence of this mysterious substance it was made to serve more comprehensive purposes. No wonder Lavoisier taunted the phlogiston chemists with the remark:[8]

> . . . that phlogiston now had to be free fire and now had to be fire combined with an earthy element; sometimes passed through the pores of vessels and sometimes was unable to do so; and was used to explain at the same time causticity and non-causticity, transparency and opacity, colour and the absence of colour . . .

Professor Butterfield comments:[9]

> . . . the last two decades of the eighteenth century give one of the most spectacular proofs in history of the fact that able men who had the truth under their very noses, and possessed all the ingredients for the solution of the problem – the very men who had actually made the strategic discoveries – were incapacitated by the phlogiston theory from realising the implications of their own work.

For the sceptic or indeed to anyone prepared to step out of the circle of Darwinian belief, it is not hard to find inversions of common sense in modern evolutionary thought which are strikingly reminiscent of the mental gymnastics of the phlogiston chemists or the medieval astronomers.

To the sceptic, the proposition that the genetic programmes of higher organisms, consisting of something close to a thousand million bits of information, equivalent to the sequence of letters in a small library of one thousand volumes,* containing in encoded form countless thousands of intricate algorithms controlling, specifying and ordering the growth and development of billions and billions of cells into the form of a complex organism, were composed by a purely random process is simply an affront to reason. But to the Darwinist the idea is accepted without a ripple of doubt – the paradigm takes precedence!

---

*The total amount of information in the genomes of higher organisms is unknown but even if only ten per cent of the DNA is informational, the problem is the same. Moreover as discussed in Chapter 14 there is a growing likelihood that the genome may contain even more than one thousand million bits of information.

Again, in the context of the almost mathematically perfect isolation of different groups of organisms at a molecular level, the Darwinist, instead of questioning the orthodox framework as common sense would seem to dictate, attempts of justifying his position by *ad hoc* proposals, molecular clocks and such, which to the sceptic are self-apparent rationalizations to neutralize what is, on the face of it, hostile evidence.

Similarly, the sorts of scenarios conjured up by evolutionary biologists to bridge the great divisions of nature, those strange realms of 'pro-avis' or the 'proto-cell' which are so utterly unrealistic to the sceptic, are often viewed by the believer as further powerful confirmatory evidence of the truth of the paradigm.

Evolutionary thought today provides many other instances where the priority of the paradigm takes precedence over common sense. Take the response by specialists in pre-biotic evolution to the implications of the shrinking time available for the origin of the cell. As mentioned in Chapter Eleven, over the past decade the estimates of the time when life first occurred on the planet have moved closer and closer to the formation of the Earth's crust. A span of time which was once measured in thousands of millions of years has now shrunk to a few hundred million at the most. The recent discovery of blue green algae in rocks nearly 3.5×10^9 years old leaves a gap of perhaps 400×10^6 years between the formation of the oceans and the appearance of life. It is beginning to look as though simple life appeared as soon as the surface waters were sufficiently plentiful and cool enough to support it. On top of this, evidence from the earliest sedimentary rocks gives no indication of a supposed primeval soup.

One might have expected, considering the great difficulty in visualizing how life might have arisen as a result of simple random processes, that the ever-shrinking time available at the roulette wheel would have caused at least a ripple of doubt in the mind of even the most earnest believer. But, on the contrary, Carl Sagan in a recent *Scientific American* article takes the shrinking time available as evidence that life is *probable*![10]

> Thus the time available for the origin of life seems to have been short, a few hundred million years at the most. Since life originated on the earth, we have additional evidence that the origin of life has a high probability.

Thus again the paradigm prevails and the holistic illusion is created that every single fact of biology irrefutably supports the Darwinian thesis. Hence, even evidence that is to all common sense hostile to the traditional picture is rendered invisible by unjustified assumptions.

Of course, the triumph is only psychological and subjective. The rationalizations are unconvincing to anyone not emotionally committed to the defence of Darwinian theory. To an outsider from the community of belief, they merely tend to emphasize the metaphysical nature of evolutionary claims and the lack of any sort of rational or empirical basis.

The anti-evolutionary thesis argued in this book, the idea that life might be fundamentally a discontinuous phenomenon, runs counter to the whole thrust of modern biological thought. The infusion with the spirit of continuity has been so prolonged and so deeply imbibed that for most biologists it has become quite literally inconceivable that life might not be a continuous phenomenon. Like the centrality of the Earth in medieval astronomy, the principle of continuity has come to be considered by most biologists as a necessary law of nature. It is unthinkable that it might not hold. To question it is an offence to all our basic intuitions about the nature of biological reality.

In fact, of course, the principle of continuity, however much it may appear an unbreakable axiom, is not a necessary law of nature. The axiom has never been proved and there is nothing in all the realm of biology, nor in the more fundamental realm of physics, which calls for the continuity of life on earth to be a necessary law of nature.

Whatever the initial source of its appeal, the concept of the continuity of nature has always suffered the enormous drawback in that at no time throughout the whole history of Western thought, from the first glimmerings of the idea on Ionia, through its theological phase in the eighteenth century, right up to its latest manifestation in twentieth century Darwinian thought, has it been possible to provide any direct observation or empirical evidence in its support. Put simply, no one has ever observed the interconnecting continuum of functional forms linking all known past and present species of life. The concept of the continuity of nature has existed in the mind of man, *never* in the facts of nature. In a very real sense, therefore, advocacy of the doctrine of continuity has always necessitated a retreat from pure empiricism, and contrary to what is widely assumed by evolutionary biologists today, it has always been the anti-evolutionists, not the evolutionists, in the scientific community who have stuck rigidly to the facts and adhered to a more strictly empirical approach.

Even in classical times Aristotle's opposition to the evolutionism of the pre-Socratics was based on his acute observation of nature and his appreciation of the facts of biology. It was, again, the actual facts that led Linnaeus, Cuvier and most of the professional biologists of the seventeenth, eighteenth and early nineteenth centuries to favour a discontinuous view of nature. As Lovejoy comments:[11]

> It was on the whole the former tendency (the tendency to perceive nature in discontinuous terms) that prevailed in early modern biology. In spite of the violent reaction of the astronomy, physics, and metaphysics of the Renaissance against the Aristotelian influence, in biology the doctrine of natural species continued to be potent – largely, no doubt, because it seemed to be supported by observation.

When Voltaire expressed his opposition to the notion of the great chain of being it was because the facts spoke for discontinuity. Lovejoy explains:[12]

> Voltaire had once, indeed, he tells us, been fascinated by the idea of the Scale of Being. . . . "When I first read Plato and came upon this gradation of beings which rises from the lightest atom to the Supreme Being. I was struck with admiration. But when I looked at it closely, the great phantom vanished, as in former times all apparitions were wont to vanish at cock-crow." . . . Voltaire's criticism is that any man who will give the slightest attention to the facts will see at once the falsity of the supposition that "nature makes no leaps."

Theology and philosophy may well have asked for continuity, but observation pointed to discontinuity. Nature refused to conform – the great chain of being was broken.

Cuvier's ridicule of the evolutionism of Lamarck was based on hard-headed empiricism, on the obvious discontinuities of nature. It was Lamarck who was retreating from the facts. When, half a century later, Agassiz referred to the notion of continuity in its new Darwinian guise as "a phantom" he was speaking as a true empiricist. It was Darwin the evolutionist who was retreating from the facts.

It was again the same basic contradiction between observation – which spoke for discontinuity – and the idea of evolution by natural selection – which demanded continuity of nature – that lay at the heart of Darwin's angst in the *Origin*. The idea that it was the opponents of evolution who were blinded by the error of *a priorism* is one of the great myths of twentieth-century biology. If anyone was blinded, it was the seekers after the phantom of continuity. How could it be otherwise when they admitted as did Darwin himself that the crucial evidence in the form of connecting links was emphatically absent? Can we accuse anti-evolutionists like Agassiz of "looking down the wrong road", a phrase used recently by Mayr, when the evolutionists themselves conceded that in the last analysis nature provided no direct empirical support for their views?[13] If the evolutionists were "looking down the right road" it was certainly not a road derived directly from the facts of nature.

Undoubtedly, one of the major factors which contribute to the immense appeal of the Darwinian framework is that, with all its deficiencies, the Darwinian model is still the only model of evolution ever proposed which invokes well-understood physical and natural processes as the causal agencies of evolutionary change. Creationist theories invoke frankly supernatural causes, the Lamarckian model is incompatible with the modern understanding of heredity, and no case has ever been observed of the inheritance of acquired characteristics; and saltational models of evolution can never be subject to any sort of empirical confirmation. Darwinism remains, therefore, the only truly scientific theory of evolution. It was the lack of any obvious scientific alternative which was one of its great attractions in the nineteenth century and has remained one of its enduring strengths ever since 1859. Reject Darwinism and there is, in effect, no scientific theory of evolution.

There is still a possibility that living systems could possess some novel, unknown property or characteristic which might conceivably have played a role in evolution. Who would have believed until a few years ago that migrating birds can sense the magnetic field of the Earth? We still have no idea how this is done, nor has anyone any idea which cells in the bird are responsive to those forces. Or who would have imagined that the genes of higher organisms would be split into non-continuous sections in the DNA? Again, no one has any idea what the function of this extraordinarily complex arrangement could be. There are many biological phenomena, particularly in the field of embryology and morphogenesis, which have not been explained in terms of modern biochemical and physiological concepts. The brain is another area which is still largely mysterious. Basic mental processes such as memory are only poorly understood. The nature of self-awareness is entirely enigmatic. And outside the realm of biology there are perfectly natural processes which have not been adequately explained. Just how the perfect hexagonal symmetry of a snowflake arises is still baffling to crystallographers. No one can be sure that all the fundamental forces of nature have been identified, nor would any physicist be so bold as to assert that the basic structure of matter is understood.

Such speculations do not, however, provide anything other than the vaguest possibility that a naturalistic alternative to the Darwinian paradigm may be possible sometime in the future.

In the case of the theories of phlogiston and the Ptolemaic cosmology, it was more than anything else the absence of conceivable alternatives which guaranteed their continued defence, even when this necessitated increasingly implausible rationalizations. The final abandonment of a theory has invariably required the development of an alternative. As Kuhn points out:[14]

> . . . a scientific theory is declared invalid only if an alternative candidate is available to take its place. No process yet disclosed by the historical study of scientific development at all resembles the methodological stereotype of falsification by direct comparison with nature . . . the act of judgment that leads scientists to reject a previously accepted theory is always based upon more than a comparison of that theory with the world. The decision to reject one paradigm is always simultaneously the decision to accept another, and the judgment leading to that decision involves the comparison of both paradigms with nature and with each other

The crisis in medieval astronomy was only resolved with the advent of Copernicus and the new heliocentric theory of the heavens, and in eighteenth century chemistry by Lavoisier and the true theory of combustion.

Whether the Kuhnian view of the role of and the priority of paradigms is right, it certainly provides a satisfying explanation of why even in the face of what are "disproofs", Darwinian concepts continue to dominate so much of biological thought today. Consequently, biologists wishing to operate within a scientific framework, even those only too well aware of the seriousness of the problems, have no alternative at present but to continue to subscribe to the

Darwinian world view. It seems more than likely that, given the need for and the priority of paradigms in science, the philosophy of Darwinism will continue to dominate biology even if more by default than by merit; and that until a convincing alternative is developed the many problems and anomalies will remain unexplained and the crisis unresolved. The lack of any scientifically acceptable competitor leaves evolutionary biology in a state of crisis analogous to the crisis in medieval astronomy when, although the Ptolemaic system was admitted to be a monstrosity, the lack of any conceivable alternative imprisoned the science for centuries within the same circle of belief.

Whatever view we wish to take of the current status of Darwinian theory, whatever the reasons might be for its undoubted appeal, whether we wish to view it as being in a classic state of crisis as described by Kuhn, there can be no doubt that after a century of intensive effort biologists have failed to validate it in any significant sense. The fact remains that nature has not been reduced to the continuum that the Darwinian model demands, nor has the credibility of chance as the creative agency of life been secured.

The failure to validate the Darwinian model has implications which reach far beyond biology. It was the overriding relevance to fields far removed from biology that made the Darwinian revolution in the nineteenth century so much more significant than other revolutions in scientific thought. In the century since 1859 the Darwinian model of nature has come to influence every aspect of modern thought. As Ernst Mayr remarks:[15]

> Einstein's theory of relativity, or Heisenberg's of statistical prediction, could hardly have had any effect on anybody's personal beliefs. The Copernican revolution and Newton's worldview required some revision of traditional beliefs. None of these physical theories, however, raised as many new questions concerning religion and ethics as did Darwin's theory of evolution through natural selection.

The entire scientific ethos and philosophy of modern western man is based to a large extent upon the central claim of Darwinian theory that humanity was not born by the creative intentions of a deity but by a completely mindless trial and error selection of random molecular patterns. The cultural importance of evolution theory is therefore immeasurable, forming as it does the centrepiece, the crowning achievement, of the naturalistic view of the world, the final triumph of the secular thesis which since the end of the middle ages has displaced the old naive cosmology of Genesis from the western mind.

The twentieth century would be incomprehensible without the Darwinian revolution. The social and political currents which have swept the world in the past eighty years would have been impossible without its intellectual sanction. It is ironic to recall that it was the increasingly secular outlook in the nineteenth century which initially eased the way for the acceptance of evolution, while today it is perhaps the Darwinian view of nature more than any other that is responsible for the agnostic and sceptical outlook of the twentieth century. What was once a deduction from materialism has today become its foundation.

The influence of evolutionary theory on fields far removed from biology is one of the most spectacular examples in history of how a highly speculative idea for which there is no really hard scientific evidence can come to fashion the thinking of a whole society and dominate the outlook of an age. Considering its historic significance and the social and moral transformation it caused in western thought, one might have hoped that Darwinian theory was capable of a complete, comprehensive and entirely plausible explanation for all biological phenomena from the origin of life on through all its diverse manifestations up to, and including, the intellect of man. That it is neither fully plausible, nor comprehensive, is deeply troubling. One might have expected that a theory of such cardinal importance, a theory that literally changed the world, would have been something more than metaphysics, something more than a myth.

Ultimately the Darwinian theory of evolution is no more nor less than the great cosmogenic myth of the twentieth century. Like the Genesis based cosmology which it replaced, and like the creation myths of ancient man, it satisfies the same deep psychological need for an all embracing explanation for the origin of the world which has motivated all the cosmogenic myth makers of the past, from the shamans of primitive peoples to the ideologues of the medieval church.

The truth is that despite the prestige of evolutionary theory and the tremendous intellectual effort directed towards reducing living systems to the confines of Darwinian thought, nature refuses to be imprisoned. In the final analysis we still know very little about how new forms of life arise. The "mystery of mysteries" – the origin of new beings on earth – is still largely as enigmatic as when Darwin set sail on the *Beagle*.

NOTES

1. Mayr, E. (1963) *Animal Species and Evolution*, Harvard University Press, Cambridge, Mass, p 586.
2. Lovejoy, A. D. (1961) *The Great Chain of Being*, Harvard University Press, Cambridge, Mass. p 233
3. Quoted in Lovejoy, op cit
4. Kuhn, T. S. (1970) *The Sructure of Scientific Revolutions*, 2nd ed, University of Chicago Press, Chicago
5. Fayerabend, P. (1965) "Problems in Empiricism", in *Beyond the Edge of Certainty*, ed E. G. Colodny, pp 145–260.
6. Kuhn, op cit, see Chapter Five.
7. ibid, p 69.
8. Butterfield, H. (1957) *Origins of Modern Science*, G. Bell & Sons Ltd, London, p 199
9. ibid.
10. Sagan, C. (1975) "The Search for Extraterrestrial Intelligence", *Scientific American*, 232(5): 80–89, p 82.
11. Lovejoy, op cit, pp 227–28.
12. ibid, p 252–53.
13. Mayr, E. (1972) "The Nature of the Darwinian Revolution", *Science*, 176: 981–989, see pp 987–88.
14. Kuhn, op cit, p 77.
15. Mayr (1972) op cit, p 988.

INDEX

Agassiz, Louis,
on classification 132
on continuity 94, 139
on nature/God relationships 20
aircraft evolution 316, 317
Alvin, the 159
Amadon, D.,
honeycreeper study 84
amino acids 234
elongation of chain 246
modification of protein, role in 321,
322
structure and function 235–8
amniotic egg, evolution of 218,
219
Anaximander of Miletus,
evolutionary philosophy of 39
angiosperms, the 163
Araeocelis, diagram of 170
Archaeopteryx 57, 58, 168, 208
flight theory 205
as intermediate form 175–8
Arctosa perita, altruistic behaviour in
222, 223
Aristotle 19, 290, 354
and biological classification 122, 123,
131, 140
Atkins, L.,
on lizard speciation 84
automata 269, 316, 337, 338
Axelrod, D.,
on angiosperms 163

bacteria,
characteristics 250
flagellum, the 223–5
methanogenic 287, 288
Barnes, Robert,
on fossil record 163
bat, evolution of 213–16
Beagle, the 17
significance of to Darwin 25
behaviour patterns 222, 223

Berg, Howard,
on flagellum 224, 225
Bertalanffy, L.,
on behaviour patterns 222
biochemistry,
basic structures 234–48
comparative 274–306
discoveries, recent 233
birds,
character traits 106, 107
evolution of 202–13
feather, the 202–7
flight theories 204–8
lung, parabronchi of 211
lung and respiratory system in 210–13
brain, the, forest analogy 330, 331
Bristow, William S.,
on behaviour patterns 222, 223
Bronn, H. G.,
on intermediate forms 103
Burgess Shale collection 161, 162, 187
Butterfield, H.,
on phlogiston theory 351
carnivorous plants 226
Carter, G. S.,
on inadequate explanations 153
catastrophe, theory of 22–4, 26, 35
cell, the *see also* individual components,
autodestruction 266, 267
automaton analogy 269
biochemical design, basic 250
magnified 328–30
mutation 267, 299
origin, hypothetical 251, 260, 263–71
self-replication 247, 248, 269, 329, 335
translational mechanism *see* separate
entry
cell translational mechanism *see also*
genetic code, and protein synthetic
apparatus, 243–5
evolution, hypothetical 264–8
Challenger, H.M.S. 159
Chalmers, T.,
and Noah's flood 23

chance *see also* random selection 308–24
Chandler, Asa,
  on parasite evolution 221
change, limits to 64, 65, 91
  language analogy 88–90
  machine analogy 90, 91
Chapman, R. F.,
  on alimentary tract development 149
character traits 117
  birds 106, 107
  invariant 134, 135
  mammals 105, 106
chloroplast, the 335
Cilium, the 107–9
cladism,
  bias of 137–40
  characteristics 128, 129
  cladograms 129, 130
  fossil taxa 165
  hierarchy 131
  opposition to 138, 139
classification *see also* cladism, and
    hierarchy, 119–40
  ambiguity 65
  Aristotle's 122, 123
  bias 138
  character traits 105–7, 117, 134, 135
  definitions 50, 51, 122
  evolutionary trees 126, 127
  molecular studies, comparative 282–6
  non-hierarchic systems 131
  phenetic 125, 128
  typological 94–8, 132, 133
  Venn diagrams 119
  vertebrates 120
Cloud, Preston,
  on fossil record 187, 188
coelacanth, the 157
  discovery of *Latimeria* 179, 180
Coleman, William,
  on science/religion conflict 21, 100
  on typology 18
comparative anatomy *see also* homology,
    48–52
  contradictions in 155
  hierarchy 50–2
  homologous and analogous
    resemblances 48–52, 153
Conan Doyle, Arthur,
  on circumstantial evidence 155
  on missing links 158

convergence, phenomenon of 178, 180,
    181
coral reef formation 26, 27
Coryanthes, pollination mechanism of
    225, 226
creationism *see also* Genesis, 52, 53
  geographical variation challenge 46
  homology 143, 144
Crick, Frances,
  on cell origin 265, 271
Crystal, R. N.,
  on behaviour patterns 223
Crytocleidus oxoniensis, diagram of 170
Cyrtopholis portoricae, altruistic
    behaviour in 223
cytochrome C,
  diversity pattern 294–6
  percentage sequence differences
    278–82
Cuvier, Georges,
  on catastrophism 22
  on change, limits to 91, 213
  religion, attitude to 100
  on typology 18, 99, 101–3, 117
Cynodictis, diagram of 171

Darwin, Charles,
  Beagle, significance of 25
  biographical details 24, 25
  doubts 67, 70, 77, 134, 326
  evolution theory, dawning of 28–35,
    42, 43
  Joseph Hooker, letter to 34
  Malthus, influence of 42, 44
  personal character 65
  religious convictions 53–5
"Darwin's Finches" 30–2, 61
Dawkins, Richard,
  on Darwinian theory 75, 76
  on random selection 324
dawn rocks, the 261
De Beer, Gavin,
  on homology 146–7, 149, 151
Dewar, D.,
  on hypothetical reconstructions 217,
    218
Dimorphodon, diagram of 168
Diver, Cyril,
  on natural selection 79
DNA (deoxyribonucleic acid),

discovery of 233
function 239
information storage capacity 334,
    336, 338
interspecies variation 287, 288
replication 245, 247, 248, 338
structure 239–41
dragonfly, mating of 219, 220
Drake, Frank,
    on extraterrestrial life 253
Duncker, H. R.,
    on bird evolution 212
Durham, Wyatt,
    on fossil record 190
Dyson, Freeman,
    on extraterrestrial technology 258–60

Ediacara Hills collection 161, 162, 187
Eiseley, Loren
    on Darwin's self-doubt 69
Eldredge, Niles,
    on saltation 192–4
    on social context 70
elephant, evolution of 184, 185
embryogenesis, and homology 145–9
Empedocles, selectionist theory of 39
enzymes 238
    structure and function 334
Eohippus 182–4, 186, 191
essentialism 19
Euparkeria, diagram of 168
European gull, speciation in 81, 82
Eusthenopteron, diagram of 167
Evidences 19, 20
evolution theory *see also* macroevolution
    early exponents 37, 39
    Lamarckian 41
    origins of 28–35
    pre-Socratic materialistic 39, 40
    saltational 192–5
    scientific status 71–3
    social context 70, 71
    social/scientific acceptance of 74–6
    substantiation issues 55–7, 75, 76, 81
    tree of life 38
    vitalistic 41
Ewens, W. J.,
    on mutation rates 298
extinction, role of 136
eye, the 61, 62, 147, 332, 333

Fabre, Henri,
    on behaviour patterns 222
Fayerabend, Paul,
    on social acceptance 76, 77, 348
feather, evolution of 202–7
fixity of species *see* typology
flagellum, bacterial 223–5
    rotary motor of 224
flight,
    arboreal theory of 204–8
    cursorial theory of 205–7
Forey, Peter,
    on coelacanth discovery 179
fossil record, the 52, 57, 157–95
    adequacy of 189, 190
    amphibian 164
    Burgess Shale collection 161
    and cranical endocasts 177, 180, 181
    discontinuous nature of, examples
        166–72
    Ediacara Hills collection 161, 162
    gaps, explanations of 186
    imperfections in 57, 188–92
    insufficient search of 160, 186–8, 191
    intermediate forms in 101–4, 162–5
    invertebrate 162, 163
    Lipalian gap, the 187, 188
    plant 163
    in pre-Cambrian strata 163, 187,
        188
    present character of 164, 165
    and saltational evolution 189, 190
    and soft anatomy 177–80, 182
    vertebrate 164, 173
fowl, domestic,
    and pleiotropic genes, effects of
        149–51
Frey-Wyssling, A.,
    on *Cilia* 108
fruitfly, Hawaiian (*Drosophila*),
    speciation in 82, 83, 93, 193

Galapagos Islands 28–35
    characteristics 28, 29
    finches of 31–2
    and variation of species 30–5, 45
Galeopithecus, and flight evolution 214,
    215
gaps *see also* intermediate forms, 164,
    165, 199–230

explanations for 186–95, 346, 347
  organic/inorganic 347
general theory, Darwin's *see*
  macroevolution
genes *see also* genetic code,
  and homologous structures, role in
    149
  overlapping 336, 337
  pleiotropy 149–51
  quantity 331, 332
  split 331, 332
  structure and function 240, 242
  substitution rate 60
Genesis *see also* creationist theory, 20–3,
  35
genetic code *see also* cell translational
  system, and protein synthetic
  mechanism, 108, 109, 243–5
genetic drift 86, 297–301
geographical variation of species, 27,
  28, 45, 46
  in Galapagos Archipelago 29–35
geology *see also* uniformitarianism,
  macro phenomena 87, 88
Geology, Principles of 24, 25, 99
George, Wilma,
  on cladism 139
Goldschmidt, R.,
  on "hopeful monster" 230
Gompez, Theodore,
  on natural selection 39, 40
Gorman, G. C.,
  on speciation in lizards 84
Gould, S. J.,
  on anti-evolutionism 104, 105
  on hypothetical reconstructions 228,
    229
  on imperfection 327
  on saltation 192–4
  on social context 70
Gruber, Howard,
  on classification 134
  on saltationism 58, 59
Guardian Weekly, The
  on missing links 194

haemoglobin,
  amino acid mutations in 323
  percentage sequence differences 284,
    286

variations and comparison 288, 289
Halitherium 57
  diagram of 171
Hallucigensia 161, 162
Hardy, Alister,
  on homology 151
Hartley, Brian,
  on protein modification 321
Heilman, Gerhard,
  on bird evolution 204, 205, 208
hierarchic classification 50–2, 119–25,
  131–9
  acceptance and persistence of 123–5
  Aristotle's 122, 123
  characteristics 119, 121, 122
  fossil taxa, inclusion of 165
  Haekel's tree 125
  interpretation ambiguity 131, 132,
    134–7, 139
  and overlapping classification,
    comparison 122
  transport example 121
Hipparion 57
Holmes, Sherlock,
  on circumstantial evidence 155
homology, phenomenon of 48–52, 65,
  142–55
  and embryogenesis 145–9, 154
  evolutionary interpretation of 143–5,
    151, 154, 155
  explanation, lack of 151, 154
  and genes, role of 149–51
  mistaken identification of 153, 154
  pentadactyl design, the 142–4, 151–3
  rudimentary organs 50
honeycreeper, Hawaiian, speciation in 84
"hopeful monster", the 230
horse, evolution of 93, 182–6, 191
  significance of 185, 186
Hoyle, Fred,
  on chance 323
  on panspermia 271
Hull, D. L.,
  on typology 101, 104
Hume, D.,
  on machine analogies 339, 340
  on random selection 40
Huxley, Julian,
  on Darwinian theory 74, 76, 226
Huxley, T. H.,
  on primate evolution 116

on saltation 192
on uniformitarianism 72
Huygens, Christianus,
    on extraterrestrial life 252
Hydra, the 346
Hylonomus, diagram of 169
hyrax, diagram of 171

Ibalia, altruistic behaviour in 223
Icaronycteris, diagram of 168
ichthyosaur 166
    diagram of 169
Ichthyostega 167
information storage,
    mechanical/biological systems 333,
        334
inheritance,
    blending 63, 64
    chemical basis of 233
    mechanism of 43, 64
    Mendelian theory 64
intelligence, human 338, 339
intermediate forms,
    absence of 57–60, 77
    absence of, explanations 186–95
    examples of 109, 110, 175–80, 292–4
    extinction, role of 136
    fossil record of *see* separate entry
    hypothetical reconstruction of *see*
        separate entry
    physical/biological worlds 249, 250
    quantity 172, 174
    typological view of 96–8, 100
intermediate forms, reconstruction of 56
    difficulties in 201, 202
    imagination factor 227–9, 250, 251

Janusch, John Beuttner,
    on primate evolution 116
Jenkins, F.,
    on blending inheritance 63
Jepson, G. L.,
    on bat evolution 215, 216
Jukes, T. H.,
    on aircraft evolution 316, 317

Kettlewell, Bernard,
    peppered moth, experiments on
        79–81

Koestler, Arthur,
    and "Beyond Reductionism" 327, 328
Kuhn, Thomas,
    on appeal of theory 348, 349, 356, 357

Lamarck, J.,
    evolution theory 41
life, extraterrestrial *see also* Mars, 252,
    253
    evidence of 260
    "Project OZMA" 253, 258
    significance of 252, 255, 260
    and technology, detection of 258–60
linguistics,
    and chance, analogy 308–12
    and change, limits to 88–90
    evolution of 199
Linnaeus, Carl,
    Systema Naturae 123
    and typology 99
Lipalian gap, the 187, 188
liver fluke, the 221
Lloyd, F. E.,
    on carnivorous plants 226
Lost World 158
Lovejoy, A.,
    on continuity 354
    and hierarchy 124
    on missing links 345, 346
Lowell, Percival,
    on Martian canals 253, 254
Lunakod 333
lungfish, the 109, 110, 180, 195, 302,
    303
Lyell, Charles,
    on age of earth 100
    and geographical variation 34
    Principles of Geology 24, 25, 99
    and uniformitarianism 24, 71, 72, 99

macroevolution *see also* natural selection
    Darwinian claim for 46–56
    extinction, role of 136
    homology *see* separate entry
    intermediate forms *see* separate entry
    and microevolution, extrapolation
        from 86–8, 92
    philosophical/religious implications
        53, 54, 66, 67, 357–9

sceptics 86, 87, 345, 347
significance, social 357–9
social context 70, 71
social/scientific acceptance of 74–7, 353–9
Malthus, Thomas,
    influence on Darwin 42
mammals,
    character traits 105, 106
    variations of 95
Mars,
    canals on 253
    characteristics 254, 257, 258, 348, 349
    life on 251–8, 261, 262
    Mariners II and IV missions to 254
    Viking mission to 255–7, 260, 335
materialistic theory, pre-Socratic 39, 40
Mayr, E.,
    Animal Species and Evolution 84, 85
    on antithetical theories 98, 99
    on classification 124–6, 132, 133
    on fixity of species 19, 47
    on "hopeful monster" 230
    on macroevolution 344
    on pleiotropy 149
    on sceptics 86, 87
    on speciation 81, 84
Medawar, Peter,
    on chance 328
Mengel, R. M.,
    wood warbler, study of 83, 84
Mesosaurus, diagram of 169
metamorphosis 220, 221
    and homology 147, 148
microevolution *see* special theory
Minsky, Marvin,
    on checker-playing 313, 314
missing links *see also* intermediate forms 345
    comparative biochemistry evidence 277
    preoccupation with 158, 160
    and seabed exploration 158, 159
molecular clock hypothesis 296–306
Monod, Jacques,
    on cell structure 250
    on creation 270
    on random mutation 43
monotremes, the 109, 110, 195
Morowitz, H. J.,
    on hypothetical cell structure 263, 264

Morris, S. C.,
    on Burgess Shale collection 161

natural selection *see also* speciation, and random selection,
    and blending inheritance 63, 64
    change, limits to 64, 65, 88–91
    islands of function 317–20
    and living fossils 301–4
    observation of 79–86, 344, 345
    premises of 42, 43
    pre-Socratic materialistic theory of 39, 40
    in protein evolution 301–6
    time factor 47, 48, 57, 60–2, 331, 352
    and variation under domestication 44, 45
Nelson, Gareth,
    on classification 138
Newell, Norman,
    on gaps 186
Noah's flood 23
Northern Arizona University,
    avian flight research 206, 207
nucleic acids *see also* DNA, RNA,
    function 238, 239, 245
    and protein interaction 248
    structure 239–41
    transcription 242, 243, 245
    types 239

Opabinia 161
opossum, the 304
Origin of Species, The, 28
    geographical variation 45, 46
    reactions to 44, 66
    religious implications 53, 54, 66, 67
    saltationism, rejection of 59
    special and general theories 44
Ostrom, John,
    on avian evolution 204–8
Owen, Richard,
    on typology 99–101

paleontology *see* fossil record
Paley, William,
    and natural theology 19, 20, 316
    and watch analogy 316, 339, 341

panspermia 271
Pattee, H. H.,
  on natural selection 314, 315
Patterson, Colin,
  on classification 138, 139
pentadactyl design, the 142–4, 151–3
peppered moth, natural selection
  experiments 79–81
Pepsis marginata, altruistic behaviour in
  223
Peripatus 110
Petrunkevitch, A,
  on behaviour patterns 223
phlogiston theory
  tenacity of 348–51, 356
Pictet, F. J.,
  on intermediate forms 103, 104
Pioneer 10 251
Platnick, Norman,
  on classification 138
Poganophora 159–61
Pompilus plumbeus, altruistic behaviour
  in 222, 223
prebiotic soup 260, 261, 323
  oxygen/ultraviolet conundrum 261,
  262
  time factor 262, 263
primate evolution 116
pro-avis 200, 205, 207
Protein Structure and Function, Atlas of
  278–80, 284, 290
protein synthetic apparatus *see also* cell
  translational system, and genetic
  code, 245, 268, 269, 338
proteins 233, 245
  functional 320–4
  functional constraints hypothesis
  299–301
  interspecies variation, comparisons
  275–87
  junk 267
  and molecular clock hypothesis
  296–306
  multi-function of 337
  and nucleic acids interaction 248
  "statistical" and "crudely made"
  265
  structural organisation of 237
  structure and function of 234–8, 320,
  329
  synthesis of 245, 268, 269

Ptolemaic theory, tenacity of 348, 349,
  356
punctuated equilibrium, theory of
  192–5

Raibert, M. H.,
  on mechanical designs 333
random selection *see also* natural
  selection 308–24
  analogies 308–13, 317
  and design perfection 326–28, 341,
  342
  possibility, assessment of 61, 62,
  319–24, 347, 348
  sceptics 327, 328, 341, 342, 351
  time factor 331
religion,
  and macroevolution, implications of
  53, 54, 66, 67, 357
  and science, relationship 19–23, 35
Rensch, B.,
  on convergence 181
  on hypothetical reconstruction 201,
  202
rhipidistian fishes,
  as intermediate forms 178–80, 182,
  194
ribosome, the 245, 246, 338
RNA (ribonucleic acid),
  function 239
  interspecies variation 287, 288
  messenger (mRNA) 242–5
  structure 240
  transfer (tRNA) 245
roadrunner, the 207
Romanes, G. F.,
  on uniformitarianism 72
Romer, A. S.,
  on intermediate forms 292
Rudwick, Martin,
  on fossil record 191
  on typology 100–3

Sacculina, metamorphosis in 220, 221
Sagan, Carl,
  on extraterrestrial life 253, 254
  on prebiotic soup 262
saltation, theory of 58, 59, 73, 186,
  192–5, 230

and chance factor 319
and creation 270, 271
"hopeful monster" 230
Satir, Peter,
   on *Cilia* 108
Schiaparelli, Giovanni,
   on Martian canals 253
Schopf, Thomas,
   on fossil record 189
seal, diagram of 171
Seymouria, as intermediate form 176,
   177
shrew, diagram of 168
Sikes, Sylvia,
   on elephant evolution 185
Simpson, G. G.,
   on classification 65, 124, 125,
      137
   on fossil record 164, 165, 189
   on horse evolution 184
   on natural selection 317, 318
Sinopa, diagram of 170
Sirex, altruistic behaviour in 223
six-thousand-year-old earth, theory of
   21, 23, 24, 26
Smith, James,
   on bat evolution 215
Soffen, Gerald,
   on *Viking* mission 254, 255
soft biology 177–80, 182
special theory, Darwin's *see also*
   speciation 44–6
   extrapolation from 86–8, 92, 344
   validation 85, 86
speciation 105
   circular overlaps 81–3, 93
   "Darwin's finches" 30–2
   definition 46
   examples, specific 81–4
   gene sequences 83
   and isolation, geographical 84–6
   model of, theoretical 84, 85
   punctuational model of 192–5
species, definition 51, 81
Spencer, Herbert,
   on social context 70
Sprigg, R. C.,
   and Ediacara Hills discovery 161
squirrel, evolution of 213, 214
Sutherland, I. E.,
   on mechanical designs 333

symposia,
   Beyond Reductionism 327
   Mathematical Challenges to the Neo-
      Darwinian Interpretation of
      Evolution 314
   Natural Automata and Useful
      Simulations 314
   Wistar Institute 328
Systema Naturae see clasification
Stahl, Barbara,
   on coelacanth discovery 179
   on feather evolution 209
Stanley, Steven,
   on intermediate forms 157, 158, 182,
      229

taxonomy *see* classification
technology,
   and biological design, analogies
      332–42
Thompson, D'Arcy,
   on discontinuity 229
Thompson, Keith,
   on cladism 139
Thomson, J. A.,
   on aquatic vertebrate evolution 216
Tillyard, R. J.,
   on dragonfly 220
Time magazine,
   on *Viking* mission 254, 256, 257
Tinbergen, Niko,
   and peppered moth experiments 79, 80
transitional forms *see* intermediate
   forms
Trevelyan, G.,
   on religious implications 66, 67
Tribrachidium 162
Tyndall, J.,
   on uniformitarianism 72
typological model of nature 93–117
   basic premises of 94, 96, 117
   and character traits 105–7, 134, 135
   and comparative biochemistry
      274–306
   eminent proponents 99
   and morphological/molecular
      divergence 290
   rationalisation 100–5, 117
   and theology 19, 20, 35, 73, 100, 101
   tree of 133

uniformitarianism 24, 71–3
  basic premises of 27
  and coral reef formation theory 26, 27
  early exponents of 24

Valentine, James,
  on intermediate forms 187
vertebrate evolution 110–16
  embryology, early 115
  heart and aortic arches, diagrams of
    112, 114
  inconsistency 113, 116
  marine aquatic, large 216, 217
Viking mission 254–7, 260, 335
Voltaire,
  on continuity 354
Von Neumann, J.,
  on automata 316, 337
Vorzimmer, Peter,
  on inheritance theory 63, 64

Walcott, Charles Doolittle,
  on Burgess Shale discoveries 161
Wallace, Alfred Russell,
  on geographical variation 34

Waltz, David,
  on artificial intelligence 339
Wardlaw, C. W.,
  on plant evolution 226, 227
Whewell, William,
  on extraterrestrial life 252, 253
Wickramsinghe, Chandra,
  on chance 323
  on panspermia 271
Wilson, A. C.,
  on functional constraints hypothesis
    300
Wittington, H. B.,
  on Burgess Shale discovery 161
Woese, C.,
  on cell origin 265, 266
  on methanogens 287
wood warbler, speciation in 83, 84

Zeuglodon 57, 58
Zoonomia 23
Zukerkandl, E.,
  on cell origin 266
  on functional proteins 320, 321
  on living fossils 291
Zygorhiza kochi, diagram of 170